Dominic Trommer

Solar Thermal Gasification of Petroleum Coke

Dominic Trommer

Solar Thermal Gasification of Petroleum Coke

Renewable Energy for the Petroleum Industry

Südwestdeutscher Verlag für Hochschulschriften

Impressum/Imprint (nur für Deutschland/only for Germany)
Bibliografische Information der Deutschen Nationalbibliothek: Die Deutsche Nationalbibliothek verzeichnet diese Publikation in der Deutschen Nationalbibliografie; detaillierte bibliografische Daten sind im Internet über http://dnb.d-nb.de abrufbar.
Alle in diesem Buch genannten Marken und Produktnamen unterliegen warenzeichen-, marken- oder patentrechtlichem Schutz bzw. sind Warenzeichen oder eingetragene Warenzeichen der jeweiligen Inhaber. Die Wiedergabe von Marken, Produktnamen, Gebrauchsnamen, Handelsnamen, Warenbezeichnungen u.s.w. in diesem Werk berechtigt auch ohne besondere Kennzeichnung nicht zu der Annahme, dass solche Namen im Sinne der Warenzeichen- und Markenschutzgesetzgebung als frei zu betrachten wären und daher von jedermann benutzt werden dürften.

Coverbild: www.ingimage.com

Verlag: Südwestdeutscher Verlag für Hochschulschriften GmbH & Co. KG
Heinrich-Böcking-Str. 6-8, 66121 Saarbrücken, Deutschland
Telefon +49 681 37 20 271-1, Telefax +49 681 37 20 271-0
Email: info@svh-verlag.de

Approved by: Zürich, ETH, Diss., 2006

Herstellung in Deutschland (siehe letzte Seite)
ISBN: 978-3-8381-3389-8

Imprint (only for USA, GB)
Bibliographic information published by the Deutsche Nationalbibliothek: The Deutsche Nationalbibliothek lists this publication in the Deutsche Nationalbibliografie; detailed bibliographic data are available in the Internet at http://dnb.d-nb.de.
Any brand names and product names mentioned in this book are subject to trademark, brand or patent protection and are trademarks or registered trademarks of their respective holders. The use of brand names, product names, common names, trade names, product descriptions etc. even without a particular marking in this works is in no way to be construed to mean that such names may be regarded as unrestricted in respect of trademark and brand protection legislation and could thus be used by anyone.

Cover image: www.ingimage.com

Publisher: Südwestdeutscher Verlag für Hochschulschriften GmbH & Co. KG
Heinrich-Böcking-Str. 6-8, 66121 Saarbrücken, Germany
Phone +49 681 37 20 271-1, Fax +49 681 37 20 271-0
Email: info@svh-verlag.de

Printed in the U.S.A.
Printed in the U.K. by (see last page)
ISBN: 978-3-8381-3389-8

Copyright © 2012 by the author and Südwestdeutscher Verlag für Hochschulschriften GmbH & Co. KG and licensors
All rights reserved. Saarbrücken 2012

Preface

This thesis presents a fundamental study of the thermodynamics and chemical kinetics of the solar thermal steam gasification of petroleum coke. The solar thermal gasification of carbonaceous materials is a process that uses concentrated solar power as the source of high-temperature process heat. The reactants, i.e. petcoke and steam, uniquely serve as the chemical source of the products being synthesis gas, which is a mixture of hydrogen and carbon monoxide.

Gasification is a state of the art process applied in today's petroleum processing industry, converting low or negative valued solid and liquid residual feedstocks into a gaseous product. The resulting gas is easily purified from compounds that would cause pollution problems and greatly extends the use of the employed solid fuels. The solar gasification process further preserves fossil resources because the energy content of the syngas product has been upgraded by the solar input in an amount equal to the enthalpy change of the reaction. The combination of petcoke and solar energy creates a link between todays petroleum-based technology and tomorrows solar chemical technology.

The thesis is performed in the framework of a joint project between the research and development center of Petróleos de Venezuela, S.A. (PDVSA / INTEVEP), the Swiss Federal Institute of Technology (ETH) and the Centro de Investigaciones Energéticas, Medioambientales y Tecnológicos (CIEMAT). The main objective of the project is the development of the required technology for the production of high quality syngas using extra-heavy Orinoco type of crude oil or derived residues such as petroleum coke using a solar thermochemical process. The joint project comprises a second thesis that focuses on the modeling of the radiative heat transfer in chemical reacting systems.

The work presented in this thesis was was performed at the Professorship in Renewable Energy Carriers at the Swiss Federal Institute of Technology Zurich (ETH). I would like to take the opportunity and thank all the people and institutions who have made this work possible and helped me in any way to finish this thesis successfully.

I am particularly grateful to Prof. Dr. Aldo Steinfeld, who supervised my doctoral thesis during the last four years. He has provided me with an excellent research environment and has helped to accomplish my dissertation by continuous support and challenge in every part of this work.

I thank Prof. Dr. Marco Mazzotti for critically reviewing this thesis and taking over my final examination as a co-examiner.

Further, I would like to thank Philipp Haueter and Andreas Z'Graggen who have greatly contributed both with engineering and numerical skills to the success of my thesis as well as the SynPet project. I will always keep in good memory the professional and constructive ambiance during the experimental campaign an PSI's solar furnace.

I thank the master students, R. Alvarez, M. Fasciana, F. Noembrini, and R. Speck as well as the students who did a semester project, C. Brander, F. Kritter, and C. Rechsteiner for contributing to this thesis with a large experimental effort.

I am grateful to the other members of the PRE group, namely Patrick Coray, Hansmartin Friess, Elena Gálvez, Wojciech Lipinsky, Tom Melchior, Reto Müller, Jörg Petrasch, Lothar Schunk, and Peter and Vita von Zedtwitz for all the discussions of the experimental and theoretical work.

Very special thanks go to my girlfriend, Anja Bär, princess of Berlin, for her love and patience that give me pleasure everyday and have helped me to endure difficult times. Also, I would like to thank my parents, Margret and Bernhard, for supporting me with their experience, and my sister, Natalie, for the good time we had.

Zurich, August 2006 $\hspace{4cm}$ Dominic Trommer

$\LaTeX\,2_\varepsilon$

Abstract

The objectives of this thesis are the theoretical and experimental analyses of the chemical kinetics and thermodynamics of the solar thermal gasification of petroleum coke.

Equilibrium computation of the stoichiometric system of petcoke and steam at 1 bar and 1300 K result in an equimolar mixture of H_2 and CO. A 2nd-law analysis for electricity generation using the gasification products indicates the potential of doubling the specific electrical output and halving the specific CO_2 emissions vis-à-vis conventional petcoke-fired power plants.

Rate laws for pyrolysis and reactive gasification are derived: Pyrolysis is modeled as a linear combination of first order decomposition reactions. The model for the reactive gasification bases upon a Langmuir-Hinshelwood type reaction mechanism considering reversible sorption of gaseous species and irreversible reactions among adsorbed species and with molecules from the gas phase. Further, the grain model is used to account for mass transfer limitations in the solid.

The rate constants are determined experimentally by thermogravimetry in the 900-1300 K temperature interval using H_2O-CO_2-Ar mixtures and gaseous reaction products are detected by GC. The TG experiments show a different reactivity for delayed coke samples if they are gasified as received, after partial gasification in an entrained flow reactor, or after temperature-treatment above 1300 K.

Experiments with laboratory scale fluidized bed reactors featuring two different modes of heat transfer were performed. The particles in the bed were heated by direct and indirect thermal radiation from the ETH's high-flux solar simulator using transparent and opaque fluidization tubes. An effect of the heat transfer mode on the reaction kinetics was not observed and rate data complies with data from the TG experiments.

The design of a 5 kW solar thermal cavity reactor with entrained flow of gas and solids operated with a fine ground coke slurry is presented. Experiments at PSI's solar furnace yielded chemical conversion of carbon and steam up to 87% and 69%, respectively, at a solar power input in the range 3.3-6.6 kW and a coke mass flow rate in the range 1.85-4.45 g/min. Operation with a stoichiometric feed produced a gas with $H_2/CO \approx 2$ and $CO_2/CO \approx 0.3$.

A mathematical reactor model is developed based on the axial dispersion model. The axial dispersion number is obtained from residence time distribution measurements. The model is extended by means of radial temperature profiles extracted from a heat transfer simulation. The kinetic model is used to calculate the reactor performance and the modeled data is compared with the experimental campaign. Mass transfer in the gas phase was analyzed and found to have no effect for the used particle size and temperatures.

The research presented in this work serves as a fundamental reaction kinetics study that can be applied for the design of solar thermal gasification reactors.

Zusammenfassung

Die vorliegende Doktorarbeit befasst sich mit der theoretischen und experimentellen Analyse der Reaktionskinetik sowie mit der Thermodynamik für die solarthermische Vergasung von Petrolkoks.

Die Gleichgewichtszusammensetzung eines stöchiometrischen Reaktionssystems bestehend aus Wasserdampf und Petrolkoks bei 1 bar und 1300 K ergibt eine äquimolare Mischung aus Wasserstoff und Kohlenmonoxid. Eine thermodynamische Analyse für die Umwandlung der Vergasungsprodukte in Elektrizität zeigt auf, dass der solare Prozess gegenüber konventionellen thermischen Kraftwerken mit Koksfeuerung die spezifische Elektrizitätsproduktion verdoppelt bei gleichzeitiger Halbierung der spezifischen Emissionen.

Geschwindigkeitsgesetze für die Pyrolyse und die oxidative H_2O-CO_2-Vergasung des Petrolkoks werden aufgestellt: Die Pyrolyse wird als Linearkombination von Zersetzungsreaktionen erster Ordnung modelliert. Das Modell der oxidativen Vergasung basiert auf einem Reaktionsmechanismus nach Langmuir-Hinshelwood, der reversible Sorptionsprozesse der gasförmigen Reaktanden und irreversible Reaktionen zwischen den adsorbierten Spezies berücksichtigt. Das 'Grain'-Modell wird verwendet, um den Stofftransport im Innern der Kokspartikel zu modellieren.

Die Geschwindigkeitskonstanten werden für den Bereich 900-1300 K experimentell bestimmt. Dazu wurden Experimente in einem Thermogravimeter mit Probeatmosphären mit unterschiedlichen Konzentrationen von Wasser und Kohlendioxid durchgeführt. Die gasförmigen Produkte wurden mittels Gaschromatographie analysiert. Die TG-Experimente ergaben verschiedene Reaktivitäten für die Vergasung von Delayed-Koks Proben. Dabei kam es darauf an, ob die Proben im Originalzustand, nach erfolgter partieller Vergasung in einem Flugstaubreaktor oder nach vorgängiger thermischer Deaktivierung bei Temperaturen oberhalb 1300 K vergast wurden, ergaben sich verschiedene Reaktionsgeschwindigkeiten.

Im Weiteren wurden Experimente mit Wirbelschichtreaktoren im Labormassstab durchgeführt. Zwei Typen von Reaktoren wurden verwendet, die sich in der Art und Weise des Energieeintrages in die Wirbelschicht unterschieden. In beiden Reaktoren wurde die Wirbelschicht durch Bestrahlung mit dem ETH High-Flux Solarsimulator beheizt. In einem Fall erfolgte der Strahlungseintrag direkt durch eine transparente Reaktorwand aus Quarz, im anderen indirekt über eine opake Keramikwand. Ein Einfluss der Art des Wärmeeintrages auf die chemische Kinetik konnte nicht festgestellt werden.

Das Design eines 5 kW Flugstaubreaktors für die solarthermische Vergasung von Petrolkoks Pulver mit Wasserdampf wird beschrieben. Experimente im Solarofen am PSI in Villigen ergaben einen Stoffumsatz von bis zu 87% für Kohlenstoff und 69% für Wasser. Dies erfolgte bei einem solaren Strahlungseintrag in der Höhe von 3.3-6.6 kW und einem Koksmassenstrom in der Höhe von 1.85-4.45 g/min. Im stöchiometrischen Betrieb ($\dot{n}_{H_2O,0} = \dot{n}_{C,0}$) bei einer Nenntemperatur von 1500 K produzierte der Reaktor ein Gas mit $H_2/CO \approx 2$ und $CO_2/CO \approx 0.3$.

Ein mathematisches Modell für einen Flugstaubreaktor wurde entwickelt, welches auf dem axialen Dispersionsmodell basiert. Die axiale Dispersionszahl wurde experimentell aus der Verweilzeitverteilung des Reaktors bestimmt. Das Dispersionsmodell beinhaltet radiale Temperaturgradienten, die von den Resultaten einer Simulation der Wärmetönung im Reaktor abgeleitet wurden. Der Stoffumsatz im Reaktor wurde mit dem Modell für die chemische Kinetik berechnet und mit den experimentellen Daten aus der Versuchskampagne mit dem 5 kW Pilotreaktor validiert. Die Analyse des Stofftransportes in der Gasphase ergab, dass dieser unter den gegebenen Bedingungen keinen limitierenden Einfluss ausbt.

Die vorliegende Arbeit behandelt die reaktionskinetischen und thermodynamischen Grundlagen für den Bau und den Betrieb von solarthermischen Vergasungsreaktoren.

Contents

Preface	i
Abstract	iii
Zusammenfassung	iv
Nomenclature	xi

1 Introduction   1
   1.1 Motivation 1
   1.2 Problem statement 2

2 Background and literature review   5
   2.1 Chemistry of gasification 5
      2.1.1 Pyrolysis 6
      2.1.2 H_2O-CO_2 gasification 7
   2.2 Survey of gasification techniques 8
      2.2.1 Conventional gasification processes 8
      2.2.2 Gasification with concentrated solar power 12
      2.2.3 Comparison 15
   2.3 Feedstock characterization 15
      2.3.1 Petroleum coke 15
      2.3.2 Chemical composition 18
      2.3.3 Sample preparation and morphology 20

3 Thermodynamics   23

	3.1	Equilibrium composition	23
	3.2	Enthalpy of reaction	27
	3.3	Second law analysis	27
		3.3.1 System setup	28
		3.3.2 Results	28
	3.4	Summary and conclusions	32

4 Reaction kinetics   33

	4.1	Heterogeneous gas-solid reactions	33
		4.1.1 Structural models for the coke particle	35
		4.1.2 Mass transfer model for the gas phase	40
	4.2	Experimental systems for kinetic analyses	45
		4.2.1 Differential and integral reactors	45
		4.2.2 Thermogravimetry	46
	4.3	Reaction mechanisms	47
		4.3.1 Pyrolysis	47
		4.3.2 H_2O-CO_2 gasification	48

5 Rate data from thermogravimetry   55

	5.1	Experimental	55
		5.1.1 Setup: Netzsch STA 409 and Varian Micro GC	55
		5.1.2 Experimental procedure	58
	5.2	Results	61
		5.2.1 Pyrolysis	61
		5.2.2 H_2O-CO_2 gasification	63
	5.3	Kinetic analyses	75
		5.3.1 Pyrolysis	75
		5.3.2 H_2O-CO_2 gasification	77
	5.4	Summary and conclusions	79

6 Rate constants from integral reactors   83

	6.1	DFB reactor	84
		6.1.1 Reactor setup	84

		6.1.2	Experimental procedure	85
		6.1.3	Results	87
	6.2	IFB reactor		95
		6.2.1	Reactor setup	95
		6.2.2	Experimental procedure	98
		6.2.3	Results	98
	6.3	Kinetic analysis		101
		6.3.1	Comparison of the DFB and IFB rate data	110
	6.4	Summary and conclusions		112
7	**SynPet 5 kW process reactor**			**115**
	7.1	Reactor design		115
	7.2	Experimental setup		117
	7.3	Results		119
		7.3.1	Reference experiment	120
		7.3.2	SynPet experiments at PSI's solar furnace	122
	7.4	Summary and conclusions		128
8	**SynPet reactor modeling**			**131**
	8.1	Choosing the correct model		131
	8.2	One-dimensional axial dispersion model		133
		8.2.1	Governing equations	133
		8.2.2	Numerical methods	135
	8.3	Axial dispersion number and RTD		138
	8.4	Chemical reaction		143
		8.4.1	Reactivity correction	143
		8.4.2	Mass transfer in the gas phase	146
		8.4.3	Reaction source term	148
	8.5	Radial temperature profiles		150
	8.6	Results and experimental validation		153
		8.6.1	Reference experimental run	153
		8.6.2	SynPet experiments in the PSI solar furnace	157
	8.7	Summary and conclusions		161

9	**Summary and outlook**	**165**
	9.1 Summary	166
	9.2 Outlook	168

Curriculum vitae   **191**

Nomenclature

a	mass specific surface, (m^2/g)
a_v	volume specific surface, (m^2/m^3)
A	surface, (m^2)
c	molar concentration, (mol/m^3)
C	mean solar flux concentration ratio, (−)
C$\langle * \rangle$	carbon site on the coke surface
d	thickness, diameter, (m)
D	diffusion coefficient, (m^2/s)
$\mathbf{D}$	dispersion coefficient of the axial dispersion model, (m^2/s)
E	probability density function, (−)
E_A	activation energy, (J/mol)
F	cumulative distribution function, (−)
H	fluidized bed height, (m)
I	solar irradiation, (kW/m^2)
j''	diffusion flux, (mol/m^2/s)
k'	elementary rate constant of the oxygen exchange mechanism, (units according to the rate law)
k	elementary rate constant of the extended mechanism, (units according to the rate law)
k_0	pre-exponential factor, (same units as rate constant)
k	thermal conductivity, (W/m/K)
K, a	lumped rate constants, (units according to the rate law)
L	length, (m)
m	sample mass, (g)
$\dot{m}$	mass flow rate, (g/s)
m''	mass flux, (g/m^2/s)
M	molar mass, (g/mol)
n	reaction order, (−)
n	mole number, (mol)
$\dot{n}$	molar flow rate, (mol/s)

n''	molar flux, (mol/m²/s)
N	species dependent numerator of a rate expression
N	number of control volumes, (−)
p	partial pressure, (Pa)
P	total pressure, (Pa)
q''	heat flux, (J/m²/s)
Q	heat rate, (W)
r	rate of reaction, (mol/m²/s)
r'	rate of reaction, (mol/g/s)
r''	rate of reaction, (mol/m³/s)
$\mathbf{r'}$	overall amount of pyrolysis products, (mol/g)
R	radius, (m)
$\mathcal{R}$	ideal gas constant, (J/mol/K)
SS_Y	sum of squares, (units of the addends Y)
t	time, (s)
T	absolute temperature, (K)
u	superficial velocity, (m/s)
V	volume, (m³)
w	weight fraction, (−)
W	work output, (kW)
x	H/C elemental molar ration, (−)
x	maximum conversion of a pseudo component with respect to the overall sample mass, (−)
X	chemical conversion, $X \in [0, 1]$, (−)
X_G	coke conversion due to H_2O-CO_2 gasification, $X_G \in [0, 1]$, (−)
X_P	coke conversion due to pyrolysis, $X_P \in [0, \sum_i c_i]$, (−)
y	O/C elemental molar ration, (−)
y	molar fraction, (−)

Greek symbols

α	conversion of pseudo component, $\alpha \in [0, 1]$, (−)
β	linear heating rate (ramp) of the thermobalance, (K/s)
ΔG	Gibbs free energy change, (kJ/mol)
ϵ	emissivity, (−)
ϵ_v	void fraction, (−)
η	process efficiency, (−)
η_P	particle effectiveness, (−)

μ	dynamic viscosity, (kg/m/s)
μ	predictand of the RTD, (s)
ν	stoichiometric coefficient, (−)
ρ_m	specific molar density per unit mass, (mol/kg)
ρ_v	specific molar density per unit volume, (mol/m^3)
σ	Stefan-Boltzmann constant, (W/m^2/K^4)
σ	variance, (s^{-1})
θ	fractional surface coverage, $\theta \in [0,1]$, (−)
θ	reduced time, (−)

Subscripts

B	bulk
$coke$	petroleum coke
$C.C.$	combined Brayton-Rankine cycle
e	electric
$e\!f\!f$	effective
$F.C.$	fuel cell
gr	graphite
G	gasification, gas
i, j	chemical species or reaction number
I	interface
$intr$	intrinsic
m	mass, mixture
obs	observed overall
P	pyrolysis, particle
$react$	reactivity
$R.C.$	Rankine cycle
$r.g.$	reactive gas
S	solid
tot	total
v	volume
wgs	water-gas shift
W, P, E	grid points of the finite volume method
w, e	east and west face of the control volume
0	initial, overall

Abbreviations

BET	Brunauer-Emmet-Teller
CFD	computational fluid dynamics
CPC	compound parabolic concentrator
DFB	directly irradiated fluidized bed
DTG	differential thermogravimetry
EGF	electric gain factor
E.O.	specific electric output
ETH	Swiss Federal Institute of Technology
FVM	finite volume method
GC	gas chromatograph
IFB	indirectly irradiated fluidized bed
IGCC	integrated gasification combined cycle
LHV	low heating value
MFR	mixed flow reactor
PCM	progressive-conversion model
PD	Petrozuata delayed
PFR	plug flow reactor
PLOT	porous layer open tubular
PSA	pressure swing adsorption
PSI	Paul Scherrer Institute
QFBR	quartz fluidized bed reactor
RMS	root mean square (error)
RTD	residence time distribution
SCM	shrinking core model
SOFC	solid oxide fuel cell
TG	thermogravimeter

Chapter 1

Introduction

1.1 Motivation

Anthropogenic emissions of greenhouse gases and other pollutants can be significantly reduced or even completely eliminated by substituting fossil fuels by cleaner fuels, e.g. *solar hydrogen*. The complete substitution is, evidently, a long-term goal. Strategically, it is desirable to consider midterm goals aiming at the development of hybrid solar/fossil endothermic processes in which fossil fuels are used exclusively as the chemical source for H_2 production, and concentrated solar power is used exclusively as the energy source of process heat.

An important example of such hybridization is the endothermic steam gasification of petroleum coke (petcoke) to synthesis gas (syngas). Petcoke is a major solid byproduct from the processing of heavy and extra heavy oils using delay-coking and flexicoking technology.[1] The syngas product, besides being a high-quality fluid fuel, is cleaner than its solid feedstock because its energy content has been upgraded by the solar input in an amount equal to the enthalpy change of the reaction.

The combination of petcoke and solar energy creates a link between today's petroleum-based technology and tomorrow's solar chemical technology. It also builds bridges between present and future energy economies because of the potential of solar energy to become a viable economic path once the cost of energy will account for the environmental externalities from burning

---

[1]Reserves of heavy and extra heavy crude oil in the Orinoco belt are estimated at $54.7 \cdot 10^9$ barrels. Current production rate of petcoke is $1.1 \cdot 10^4$ tons/day in the Orinoco belt [59] and $1.1 \cdot 10^5$ tons/day in the world [14] (anticipated worldwide petcoke production rate of the year 2000).

fossil fuels, such as the cost of greenhouse gas mitigation and pollution abatement. Hybrid solar/fossil processes such as the one presented in this thesis offer a viable route for fossil fuel decarbonization and create a transition path towards solar hydrogen [85, 95].

The use of the sun as a source of primary energy has several advantages: The solar energy reserve is essentially unlimited, free of charge, and its use is ecologically benign. However, solar radiation is dilute ($\approx$ 1 kW/m^2), intermittent, and unequally distributed over the surface of the earth [84]. It is therefore necessary to convert solar energy into chemical energy carriers - so called solar fuels - that can be stored long term and transported long range.

There are basically three pathways for the production of fuels with solar energy: the electrochemical, the photochemical, and the thermochemical pathway [84, 82]. The solar gasification of petcoke is an example for a thermochemical process. The basic idea is to concentrate the dilute sunlight over a small surface area with the help of parabolic mirrors and capture the radiant energy by means of a suitable reactor or receiver to obtain heat at high temperatures that can be used to run endothermic chemical reactions producing storable and transportable fuels. These solar fuels finally contain solar energy in the form of chemical energy.

1.2 Problem statement

The main objectives of this thesis are the theoretical and experimental analyses of the chemical kinetics and thermodynamics of the solar thermal gasification of petroleum coke using temperature and pressure conditions encountered in solar thermal reactors. The accomplishment of the following points is thereby of particular interest:

- A study of the chemical thermodynamics of the steam gasification of petcoke based on a simplified net reaction representing the overall chemical conversion of a carbonaceous fuel with the composition of petcoke. The results of this study include the thermodynamic equilibrium composition, the syngas yield and the enthalpy change of the gasification reaction.

- The thermodynamic study is complemented with a 2nd-law (exergy) analysis to examine technically viable routes for extracting power from the gasification products and to establish a base for comparing them

1.2. PROBLEM STATEMENT

with electricity generation using conventional power plants, especially in terms of their CO_2 mitigation potential.

- Formulation of a set of kinetic rate laws based on elementary reaction mechanisms describing reversible adsorption/desorption processes and irreversible surface chemistry. The kinetic equations are amended with the necessary mass transfer relations for heterogeneous gas-solid reaction systems to enable the calculation of gasification rates over a wide range of temperatures.

- Assessment and development of suitable experimental procedures facilitating the experimental determination of the feed stock dependent kinetic parameters including the Arrhenius-type temperature dependence for conditions prevalent in solar thermal reactors.

- Contribute to the development of a 5 kW prototype reactor accomplishing the solar thermal gasification of petroleum coke by direct irradiation in the framework of the joint project.

- Development of a mathematical model for the prediction and interpretation of the 5 kW process reactor performance features, such as amount and composition of the produced gas as well as conversion of reactants. The kinetic models to be developed offer the possibility of describing local conditions in the gasification reactor, thus representing the process that leads to the reactor outlet conditions.

The numerical reactor model consists of differential balance equations for the chemical species and includes rate expressions for chemical reactions and mass transfer. Because kinetic models provide the most detailed description of gasification reactors they require a large number of parameters that are feedstock dependent (activation energies, pre-exponential factors, etc.). Typical results of kinetic models are temperature and component concentration profiles in axial and radial directions [14].

The model further has the purpose of validating the laboratory scale experimental rate data using process conditions of the 5 kW scale and find improvements to the design of the process reactor.

Chapter 2

Background and literature review

2.1 Chemistry of gasification

The gasification of carbonaceous feedstocks is an overall process that occurs via pyrolysis reactions (thermal decomposition and devolatilization) and subsequent heterogeneous gas-solid reactions of the pyrolysis residue (coke, char) with reactive gases such as oxygen, steam, carbon dioxide, and hydrogen.

Besides carbon, the solid feedstock usually contains other elements including hydrogen, oxygen, nitrogen, sulfur, heavy metals, and trace elements in varying amounts. In a gasification process the solid feedstock is converted to a gaseous product, which can be used either as an energy source or as a raw material for the synthesis of chemicals and various fuels. The resulting gaseous product can be handled with more convenience and is easily purified from compounds that would cause pollution problems. Thus, gasification is an upgrading process that greatly extends the uses of solid fuels [14].

From a chemical point of view, gasification is essentially an incomplete combustion. In contrast to the combustion process working with excess oxygen, gasification processes operate at substoichiometric conditions with controlled oxygen supply. In an autothermal gasifier both heat is produced by internal combustion and the solid is converted to a new gaseous fuel. Dilution of the product gas with nitrogen and excess CO_2 can be avoided by substituting external heating for the internal combustion. Heat sources that apply for external heating are conventional firing systems, concentrated solar energy, or nuclear heat. However, all of the named heat sources are afflicted with a less favorable heat transfer characteristic.

As a carbonaceous feedstock passes a gasification reactor or gasifier, the following physical, chemical, and thermal processes may occur sequentially or simultaneously, depending on the reactor design and the type of feedstock [71]:

- drying,
- pyrolysis (devolatilization),
- H_2O-CO_2 reactive gasification,
- combustion.

With respect to the solar thermal steam gasification of petcoke, the drying and the combustion are omitted because petcoke contains less than 1 wt% of moisture and the reactive gas does further not contain oxygen, which is required to run the combustion reaction. The pyrolysis and H_2O-CO_2 reactive gasification steps are discussed in the following.

2.1.1 Pyrolysis

Pyrolysis is the destructive chemical decomposition induced in organic materials by heat in the absence of an oxidizing agent such as oxygen, steam, or carbon dioxide. It is an endothermic process and requires the addition of heat. Pyrolysis converts the carbonaceous feedstock into a carbon rich solid (char or coke) and liquid derivatives, oil, tar, and fuel gases such as hydrogen, methane, carbon monoxide and light hydrocarbons. The nature of the pyrolysis process and the evolving products are closely related to the operating conditions and to the composition and properties of the feedstock.

Many parameters besides the already mentioned feed composition have an effect on pyrolysis [71]:

- *Particle size*: For particles bigger than 100 μm an increase of the particle size leads to higher devolatilization times. Below 100 μm the particle size has no effect [51].

- *Heating rate*: The influence of the heating rate is difficult to measure, but it is generally believed that rapid heating prevents secondary reactions among the pyrolysis products yielding higher volatile components than slow heating. The char produced by fast heating is more reactive than char from a pyrolysis process featuring low heating rates.

2.1. CHEMISTRY OF GASIFICATION

- *Temperature*: The reactor temperature has a significant impact on both the pyrolysis product yield and the product composition. Pyrolysis can be subdivided into low temperature and high temperature processes. High temperature pyrolysis produces mainly gases, while the main products of low temperature pyrolysis are tar and heavy oil.

Despite the fact that petcoke already is a solid pyrolysis product derived from petroleum residues it still contains a certain amount of volatile matter (cf. Table 2.1 on page 19). This volatile matter can be removed in a further pyrolysis step, the so-called *calcination*. During calcination petcoke is heated in rotary kilns to temperatures up to 1400 °C to remove remanent moisture and higher boiling hydrocarbons.

2.1.2 H_2O-CO_2 gasification

Gasification of carbonaceous materials, such as petroleum coke with steam ('steam gasification') and carbon dioxide ('dry gasification'), is a heterogeneous gas-solid reaction of solid carbon and steam/carbon dioxide in the gas phase. The net process is endothermic by about 50% of the feedstock's low heating value and proceeds at above 1300 K to produce, in equilibrium, an equimolar mixture of H_2 and CO. The latter is the so called *synthesis gas* that can be used to fuel highly efficient gas turbines or as a basic commodity for the chemical synthesis of higher hydrocarbons by the Fischer-Tropsch process.

The CO/CO_2 ratio in the product gas can be further adjusted in a water-gas shift reactor increasing the H_2 content in the product gas by conversion of CO to CO_2. If, in addition, a CO_2 separation and sequestration step is added the process can be used to produce CO_2-free hydrogen [94] and gasification becomes a decarbonization process.

The key parameters controlling the rate of the oxidative gasification are temperature and partial pressures of the gaseous reactants. The mode of action is described in detail in Chapter 4.3. Due to the exponential temperature dependence defined by the Arrhenius law (cf. Equation (4.36) on page 48), the influence of temperature exceeds the effect of the reactants partial pressures.

H_2O-CO_2 gasification involves a series of chemical reactions including adsorption and desorption of gases on the coke surface, reactions among adsorbed species, and the release of carbon to the gas phase by the formation and desorption of a surface oxygen complex. The respective reactions are presented in detail in Chapter 4.3, and are not further worked out at this point.

In contrast to pyrolysis, which is essentially a unimolecular decomposition reaction, the rate of the reactive gasification step not only depends on temperature and fuel properties but also on type and concentration of a gaseous reactant at the solid surface. The reactive gasification requires therefore the transport of reactants from the bulk gas to the solid and of products from the solid to the bulk gas. In addition, the fluid dynamics inside the reactor are important because they have an effect on the mass transfer rate between the solid and the gas phase.

2.2 Survey of gasification techniques

Gasification technologies have been applied commercially for more than a century. Originally, gasification was used for the production of fuel or town gas for lighting and heating. In the first half of the last century the main interest shifted to synthesis gas for the production of chemicals and liquid motor fuels.

More recently, petroleum-derived residues have gained attention as gasification feedstocks because of a continuous decrease in the volume of conventional crudes which have to be replaced by heavier crudes. The crude oil processed in U.S. refineries over the last twenty years is characterized by a continuous increase in the gravity and sulfur content [25]. The higher volume of heavy crudes results in a higher yield of residues and a higher demand for disposal and conversion technologies. Besides technological and economical reasons, gasification is favored by environmental restrictions increasing in many countries and causing a higher demand for clean technologies.

Newer developments in the field of gasification are the integrated gasification combined cycle (IGCC) for clean and efficient electricity production based on low or negative valued feed stocks. Coproduction plants offer the simultaneous production of chemical commodities and electric energy and the integration of gasification with refineries offers new business opportunities [86, 25].

2.2.1 Conventional gasification processes

This chapter contains a survey of conventional gasification techniques. First, a set of criteria commonly used for the classification of autothermic gasifi-

2.2. SURVEY OF GASIFICATION TECHNIQUES

cation processes is presented, followed by some representative examples to elaborate on the conventional reactor technology.

Several economical, technical, and chemical aspects of the gasification of solid carbonaceous materials are summarized in the literature [14, 71, 86, 25, 96, 42, 74, 1, 23, 75]. There are more than 150 companies around the world that sell systems based on gasification concepts. Many of these are optimized for the efficient conversion of specific feedstocks such as waste, petroleum residues, coal or biomass, and operate at various scales depending on particular demands such as energy production or amount of solids to process. As a consequence, a large amount of different process implementations are available on the market.

Classification of gasification systems

Conventional gasification processes can be classified based on the method used to generate heat for the gasification reactions, on the contacting pattern of the reactants, and on the physical state of the residue removed [96].

Heat transfer mode. Based on the heat transfer a gasification process can be *autothermic* or *allothermic*. In an autothermic process, internal combustion of a part of the solid feedstock with oxygen produces the heat required to run the gasification reaction inside the reactor. This has the advantage that any losses associated with the heat transfer are avoided and the construction of the gasifier is simplified. Further, the temperature can be adjusted quickly and accurately by regulation of the oxygen content in the gasification agent. In an allothermic process, the heat required is transferred by means of a gaseous, liquid (molten slag, salts or metals), or solid heat carrier brought into direct contact with the reactants or indirectly via heat exchange surfaces.

Contacting pattern. The contacting pattern describes how the solid fuel is brought into contact with the gasification agent. Based on this, a gasification process can be classified as *fixed bed*, *fluidized bed*, or *entrained flow*, as depicted schematically in Figure 2.1. The choice of a specific pattern has an important effect on the type, rank, and size distribution of the solid fuel to use. Further, it determines the residence time, reactor temperature and pressure, and some characteristics of the produced gas.

Fixed bed gasifiers [Figure 2.1 (a)] contain a bed of lump fuel maintained at a constant depth. The coke is fed from the top end and flows countercurrent to the rising gas stream. A single particle moving through the bed passes

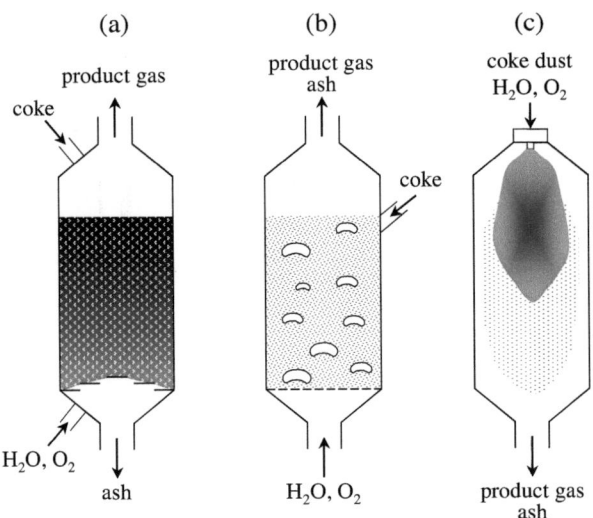

Figure 2.1: Gasifier types for autothermic gasification: (a) fixed bed, (b) fluidized bed, and (c) entrained flow gasifier.

different zones including drying and preheating, devolatilization, gasification, oxidation, and ash removal. Fixed bed gasification systems are simple, reliable and offer high efficiency with respect to coke and energy consumption.

Fluidized bed gasifiers [Figure 2.1 (b)] accept coke as grains with a size of less than $\approx$ 3 mm. The gasification agent maintains the fuel in a suspended state of continuous random motion. Reactors of this type are characterized by high mass and heat transfer rates, uniform and moderate temperatures, a high specific gasification rate and product uniformity as well as high tolerance to a wide range of feed stock compositions.

Entrained flow gasifiers [Figure 2.1 (c)] are operated with pulverized coke with particles of less than 0.12 mm. The solid fuel is entrained with the gasifying agent to react in a concurrent flow having the form of a high temperature flame. Entrained bed gasifiers accept virtually all types of carbonaceous feed stocks. Swelling and caking characteristics of the solid fuel do not affect the operability of the process as particle interaction is poor. Full entrainment of the particles requires relatively high flow rates leading in turn to small

2.2. SURVEY OF GASIFICATION TECHNIQUES

residence times. In order to obtain full conversion of the coke entrained flow reactors operate at very high temperatures (above 1800 K). Devolatilization products are released in the high temperature region and thus further cracked and oxidized. In opposition to combustion, sulfur and nitrogen contained by the solid fuel is converted to H_2S, small amounts of COS, and NH_3 rather than SO_2 and NO_x.

Ash removal. The last criterion to distinguish a gasification process is the physical state of the residue removed. Gasification reactors can operate in the *slagging* or *non-slagging* mode depending on wether the ashes liquefy or not.

Commercial gasification systems

Common gasification technologies that can serve as representative examples for a fixed and fluidized bed gasification process are the Lurgi and Winkler gasifier, respectively [14, 71, 96].

The *Lurgi dry ash* gasifier is a pressurized, dry ash, moving bed gasifier [cf. Figure 2.1 (a)]. Sized solid fuel enters the gasifier from the top through a lock hopper and moves down through the bed. From the bottom the reactor is fed with steam and oxygen reacting with the solid fuel while moving up through the bed. Ash is removed at the bottom by a rotating grate and a lock hopper. The counter flow of gas and solid results in a temperature drop in the fixed bed. High temperatures are obtained in the combustion zone at the bottom, whereas gas temperatures are lower in the drying and devolatilization zone near the top end.

The *High Temperature Winkler* gasifier consists of a refractory-lined pressure vessel equipped with a water jacket [cf. Figure 2.1 (b)]. The fuel is pressurized in a lock hopper and then pneumatically conveyed to a coal reservoir. From there, the solid drops via a gravity pipe into the fluidized bed, which is formed by particles of ash and unreacted fuel particles. The gasifier is fluidized from the bottom with either air or oxygen/steam and the temperature of the bed is kept at around 800 °C, below the ash fusion temperature. Raw syngas is passed through a cyclone to remove particulate and subsequently cooled. Recovered solids are recirculated and the ash is removed at the bottom via a discharge screw.

For the production of synthesis gas from oil residues two processes featuring an entrained flow concept are accepted worldwide: the Shell and the Texaco process, both operating under elevated pressure and without the addition of a catalyst. The Texaco process was originally designed for partial oxidation of natural gas; the Shell process was tailored from the beginning to gasification of heavy oil. The differences between the two processes are virtually negligible and involve only details of the equipment design.

The *Texaco* gasifier [96, 71, 14] consists of a pressure vessel with a refractory lining that operates at temperatures in the range of 1250 to 1450 °C and pressures of 3 MPa for power generation and up to 8 MPa for H_2 and chemical synthesis. Oxygen and steam are introduced through burners at the top of the gasifier. Solid feedstock such as coal and coke are pre-processed into a slurry by fine grinding and water addition and then pumped into the burner. Raw gas and molten ash produced during the gasification process flow out toward the bottom of the gasifier, where they are cooled and cleaned from slag ash either by water quenching or by means of a radiant cooler.

The *Shell Coal Gasification Process* [20, 19, 96, 71] is a dry-feed, oxygen-blown, entrained flow coal gasification process, which has the capability of converting virtually any coal and coke into a clean medium Btu synthesis gas. High pressure nitrogen or recycled syngas is used to pneumatically convey dried, pulverized fuel to the gasifier. The solids enter the gasifier through diametrically opposed burners, where it reacts with oxygen at temperatures exceeding 1370 °C. The gasification temperature is maintained sufficiently high to ensure that the mineral matter in the fuel is molten and smoothly flows down the gasifier wall to the slag tap. The hot syngas exiting the gasifier is quenched to below the softening point and then further cooled in the syngas cooler. Essentially all nitrogen and sulfur compounds in the feedstock are converted to their elements. The process is well suited for coke utilization and gasifies petcoke with the same conversion and sulfur removal efficiencies that are observed for coal.

2.2.2 Gasification with concentrated solar power

Gasification with concentrated solar power is a hybrid solar/fossil endothermic process in which fossil fuels are used exclusively as the chemical source for hydrogen production, and concentrated solar power is used exclusively as the energy source of process heat. A schematic of the process is shown in Figure 2.2.

The reactants being steam and a carbonaceous solid fuel such as petcoke are fed to a solar cavity reactor and gasified with high temperature solar process heat. The main products are H_2, CO, and CO_2 with compositions depending on the process conditions, such as temperature and stoichiometry

2.2. SURVEY OF GASIFICATION TECHNIQUES

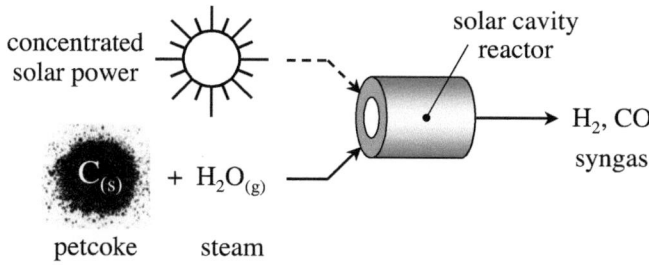

Figure 2.2: Schematic of the solar thermal gasification of petcoke.

of the feed.

Since solar gasification is performed without internal combustion as it is done in conventional gasification processes described in Chapter 2.2.1 the oxidation degree of the product gas is lower. Further, the products are not diluted with nitrogen as it happens if air is used as gasification agent. The calorific value of the products is upgraded vis-à-vis the feedstock, and the syngas quality is superior to a conventional autothermic gasification system. In addition, the solar process helps to preserve carbon resources and avoids CO_2 emissions to the atmosphere.

In analogy to the autothermic and allothermic process design of conventional gasification, the solar process supports two heat transfer modes: Efficient heat transfer to the reaction site is obtained with *direct irradiation* of the solid reactant, whereas *indirect irradiation* across an opaque reactor wall has the same drawbacks with respect to the transfer of thermal energy as observed for conventional allothermic gasification systems.

The gasification of carbonaceous materials and related reactions have been performed using concentrated solar energy in exploratory early studies with coal, oil shales, biomass, and other carbonaceous feedstocks. Experimental work dates back to the late 1970's and comprises work performed with reactor setups known from conventional gasification such as fixed bed and fluidized bed. A summary of previous work is presented in the following.

[31, 30] used a gravity feed moving bed reactor directly exposed to solar radiation in a 23 kW solar furnace. Sunlight entering the reactor through a window was focused directly on the solid fuel. Steam and CO_2 passing through the solar heated bed reacted with carbon to form a combustible product.

[90] applied a similar reactor concept, adapted for a beam down arrangement, and compared the performance of the fixed bed reactor to that of an indirectly irradiated fluidized bed reactor in a 2 kW vertical-beam solar furnace using charcoal, steam, and CO_2 as reactants. A numerical model of the moving chemical bed reactor for gasifying coal using concentrated solar radiation is proposed by [7, 8].

[6] and [4] investigated the high temperature solar pyrolysis of coal and the solar fired biomass flash pyrolysis, respectively. Powdered coal samples were exposed to concentrated solar radiation in a horizontal reactor arrangement with a quartz dome and biomass was pyrolyzed in a windowed free fall reactor.

[55, 54] developed a two stage fluidized bed hybrid coal gasification system capable of 24-hour operation. A portion of the syngas is withdrawn as product and the remaining portion acts as a heat-transfer and fluidizing medium heated by a solar receiver. Another concept for the use of a solar receiver in combination with an allothermic gasification reactor is presented by [46].

[9, 64, 38] studied the gasification reaction of various reactants in tubular reactors enclosed by a cavity type black body solar receiver. [9] used fixed and fluidized bed reactors for the pyrolysis of oil shales. [38] and [64] used transparent and opaque fluidization tubes and characterized chemical conversion and composition of the product gas at 730-950 K and 1050-1600 K, respectively.

[56] presents a study of solar thermal energy conversion by thermal storage. An infrared furnace is used as radiation source heating a mixture of coal and molten salt to 1173 K. The gasification agent is CO_2.

[24] and [44] performed laboratory scale experiments with directly irradiated fixed bed and fluidized bed reactors, respectively. Reactants are heated by direct irradiation with concentrated radiation from a solar simulator. More recently, the reaction kinetics of steam gasification of coal were investigated for a quartz tubular reactor containing a fluidized bed, and directly exposed to an external source of concentrated thermal radiation [63].

Several of the aforementioned solar reactor applications tested with small-scale prototypes were filed for a patent [29, 53, 69].

2.2.3 Comparison

The solar thermal steam gasification of petcoke provides a valuable alternative to conventional gasification processes. The advantages of supplying solar energy for process heat are four-fold:

1. The calorific value of the feedstock is upgraded and fossil fuel resources are preserved due to the exclusive use of the feedstock as the chemical source of the product gas.

2. The gaseous products are not contaminated by the byproducts of combustion, such as CO_2 and N_2 if air is used as the gasifying agent.

3. The discharge of pollutants to the environment is avoided.

4. No capital investment for an air separation unit arises compensating for the more complex reactor technology and solar radiation concentration installations.

The direct irradiation of reactants further provides an efficient means of heat transfer directly to the reaction site. Reaction kinetics are enhanced, and the limitations imposed by indirect heat transport are bypassed similar to a conventional autothermic reactor heated by internal combustion.

2.3 Feedstock characterization

This chapter outlines some of the basic feedstock properties that are important for the experimentation and modeling work of this thesis. The production of petcoke is described in Chapter 2.3.1 with special consideration of the two coke types used in this thesis. Next, the chemical composition of the cokes is presented in Chapter 2.3.2 and the sample preparation procedure is presented in Chapter 2.3.3.

2.3.1 Petroleum coke

The gasification experiments presented in this thesis are performed with petcoke being a gray to black solid consisting mainly of carbon. Two different types of petcokes are used: A *Flexicoke* and a *delayed coke* (PD coke) derived from the processing of Venezuelan extra heavy crude oil from the Petrozuata oil field located in the Orinoco belt. Manufacturer of the petcokes is the Venezuelan oil company PDVSA.

From a chemical point of view, coking can be considered as a severe thermal cracking process in which one of the end products is carbon (coke). Coking is used to convert low grade feed stocks to coke, gas, and distillates. Compared to other carbonaceous fuels such as coal, petcoke has a low reactivity under normal conditions.

Two types of coking processes are predominantly used in refineries [14, 79, 27, 15]:

1. Delayed coking is a semibatch process that uses multiple coking chambers whereof one drum is making coke from refinery residues and one drum is being decoked.

2. Fully continuous processes, such as Fluid- and Flexicoking, where hydrocarbons are coked in a fluidized bed and coke is withdrawn as a fluidized solid.

Typical feedstocks for coking units are crude oil residues obtained from the vacuum distillation. The products obtained from coking are gases, naphtha, fuel oil, gas oil, and solid coke. The formation of large quantities of coke is usually a severe problem unless the coke can be put to use. Possible options for the use of coke are combustion as a means of steam raising, gasification and stockpiling. In opposition to coal, petcoke is low in volatiles and contains a substantial amount of vanadium, nickel, and sulfur.

A considerable amount of the produced delayed coke is stabilized in an additional calcination step, which includes the elimination of water and higher boiling hydrocarbons contained in the petcoke at temperatures around 1400 °C. Under calcination conditions an expansion of the crystallite size takes place because of the cracking of side chains. At the same time the original hydrogen content of about 3.5 wt% is reduced to below 0.1 wt%. Under graphitization conditions ($\approx$ 2900 °C) good crystallized forms of petcoke are transformed into graphites that can be used as anodes for the metal reduction [14].

Delayed coking

Delayed coking is the oldest and most widely used coking process. It makes use of the fact that refinery residues can be heated above the coking point without significant coke formation as long as heaters with a sufficiently high flow rate are used.

2.3. FEEDSTOCK CHARACTERIZATION

Delayed coking units consist of a fractionator, a heater, and a pair of coke drums whereof one is on stream and one is cleaned. The bottom of the fractionator column is heated above the coking point as aforementioned and transfered to large insulated coking drums that provide long residence times ($\approx$ 24 hours) and allow the cracking reactions to proceed to completion. The heater typically operates at 480 to 500 °C and the temperature in the coke drum ranges from 415 to 450 °C at pressures from 100 to 620 kPa.

The condensation reactions that give rise to the highly aromatic coke product also tend to retain sulfur, nitrogen, and metals causing an enrichment of these elements in the solid coke. Therefore, petcoke has a high content of sulfur and heavy metals imposing an additional restriction with respect to a further use of petcoke as a fuel for combustion processes [79, 27, 15].

Fluid and Flexicoking

Fluid coking is a continuous process consisting of two vessels, a fluidized bed reactor where the feed is converted to coke and a burner for the generation of the required heat. Small coke particles produced in the process circulate between the vessels acting as a heat transfer medium. Flexicoking is a direct descendant of fluid coking and uses the same configuration as a fluid coker but includes an additional gasification step that keeps the solid inventory constant by gasification of excess coke.

The reactor contains a fluidized bed of coke particles, and steam is introduced at the bottom as fluidization agent. The feed consisting of refinery residues is injected directly into the reactor. The temperature in the coking vessel ranges from 480 to 570 °C, pressure is atmospheric and residence time is in the order of 15-30 s. Part of the incoming feed is vaporized, and part deposits on the fluidized coke particles. The material on the particle surface cracks and vaporizes to form laminated layer of dry coke. Coke is constantly withdrawn from the reactor and fed to the burner where it is partially burnt to generate the reactor heat. Excess coke from the reactor-burner circuit is fed to a fluidized bed gasifier and converted to a low heating value gas using steam and air.

Flexicoke exhibits an onion type structure. Because of the higher thermal cracking severity used in the fluid coker compared to the delayed coker, Flexicoke is much harder and denser than delayed coke. It also contains a smaller amount of volatiles and has the highest metal content of all cokes.

2.3.2 Chemical composition

The composition of the coke is expressed in terms of the proximate and ultimate analyses, which are generally used to characterize solid fuels. The respective data for PD coke and Flexicoke are listed in Tables 2.1 and 2.2.

Proximate analysis. The proximate analysis includes the determination of moisture, volatile matter, fixed carbon (by difference), and ash using methods defined by the ASTM. PD coke contains about 1% moisture, 5.8% volatiles and as little as 0.3% ash. The amount of moisture and volatiles in Flexicoke is lower because of the more severe conditions of the Flexicoking process, and the ash content is higher than that reported for PD coke.

Since for both coke types the amount of ashes is very low compared to the overall sample weight, the ash content is not included in the calculation of the chemical conversion. Moreover, the formation of an ash layer around the coke particle is excluded from the heterogeneous models presented in Chapter 4.1 as a possible source of mass transfer limitation between the gas and solid phase. Nevertheless, there is a possible catalytic effect of mineral matter present in the ash that affects the overall reactivity of the coke. Catalytic effects of certain coke components are assumed to be adequately modeled by the reaction kinetics taking into account the specific reactivity of the different coke types.

Ultimate analysis. Table 2.2 presents the ultimate analysis describing the approximate main elemental chemical composition of the coke in percent solid mass. The obviously most abundant element on a weight basis is carbon with 88 and 93% for PD coke and Flexicoke, respectively. The high carbon content is a direct consequence of the coking process converting refinery residues into a carbon rich solid, gas, and distillates. The second-most important element on a mass basis is sulfur followed by hydrogen, nitrogen, and oxygen.

The sulfur content in PD coke being 4.1% is more than twice that of Flexicoke and states a serious drawback with regard to the use of coke as a fuel for a combustion process due to the formation of SO_2.

The weight fraction of hydrogen is relatively low for both cokes but still accounts for more than 4% of the total mass in the case of PD coke vis-à-vis only 0.67% in the case of Flexicoke. Taking into account the low molar weight of hydrogen results in a molar hydrogen fraction of 34.7% for PD coke. Hydrogen is therefore the second-most frequent element behind carbon in this coke type accounting for as much as one third of all atoms.

2.3. FEEDSTOCK CHARACTERIZATION

Table 2.1: Proximate analysis for Flexicoke and PD coke (source: PDVSA).

		PD coke		Flexicoke	
		as received	dry basis	as received	dry basis
moisture	(wt%)	0.96		0.52	
ash	(wt%)	0.33	0.33	0.40	0.40
volatile	(wt%)	5.79	5.85	4.87	4.90
fixed carbon	(wt%)	92.92	93.82	94.21	94.70

Table 2.2: Approximate main elemental chemical composition (ultimate analysis), low heating value, and elemental molar ratios of H/C and O/C, for PD coke and Flexicoke.

		PD coke		Flexicoke	
		(wt%)	(mol%)	(wt%)	(mol%)
carbon		88.21	62.06	92.70	90.11
hydrogen		4.14	34.71	0.67	7.76
nitrogen		2.28	1.37	0.90	0.75
oxygen		1.46	0.77	0.92	0.67
sulfur		4.16	1.10	1.98	0.72
nickel	(ppm)	414		2 263	
vanadium	(ppm)	2 207		16 202	
sodium	(ppm)	100		326	
LHV	(kJ/kg)	35 876		32 983	
H/C	(mol/mol)	0.5581		0.0859	
O/C	(mol/mol)	0.0124		0.0074	

Petcoke further contains a substantial amount of mineral matter consisting mainly of nickel, vanadium, and sodium. Type and amount of metals in the coke strongly depends on the respective concentrations in the feedstock used for coking, as essentially all of the minerals contained in the feed are found again in the produced coke.

2.3.3 Sample preparation and morphology

Sample preparation. The petcoke used for the experiments was received in the form of raw coke coming directly from the refinery. The delayed coke consisted of coarse particles having a characteristic particle size in the order of a few centimeters and had to be ground in order to use it for the experiments. Three different grinding techniques have been applied to obtain particles with different diameters:

- A cutting mill of PSI (Villigen, Switzerland) was used to obtain a feed stock with particles in the order of 40-1000 μm.

- Ball and jet mills of ARP GmbH (Leoben, Austria) were used to obtain powder size particles with mean diameters of 3.8 and 1.3 μm, respectively.

The product of the ball and jet mills was used without further processing, the product from the cutting mill was further sieved through a tower with 80, 160, 200, 250, 355, and 500 μm sieves to obtain samples with a distinct range of particle diameters. An overview of the samples with their respective size properties is presented in Table 2.3.

Coke morphology. Figure 2.3 (a) shows a scanning electron micrograph (SEM) of PD coke with particle size 250-355 μm, magnified 200 times. The particles have sharp edges and an irregular shape originating from the grinding procedure. The solid is otherwise homogeneous and does not show agglomeration of smaller units. Also, pores are not recognizable at the applied level of magnification.

Figures 2.3 (b) and (c) show samples having the same initial particle size as (a) after gasification in the TG and DFB setups that will be presented in Chapters 6 and 5, respectively. Figure 2.3 (b) shows a particle with 21% carbon conversion. Partial gasification of the solid caused the formation of holes and and a widening of the pores across the unreacted solid. At higher conversions, the particles have a sponge-like appearance and the coherent structures in the solid decrease [cf. Figures 2.3 (c) and (d)].

2.3. FEEDSTOCK CHARACTERIZATION

Table 2.3: Coke samples used for the experiments. All numbers are lengths in μm.

coke type	mill type	mesh size	$d_{P,50}$	μ	σ
PD coke	cutting mill	< 80	26	36.6	30.1
		80–160	121	145.9	68.0
		160–200	212	249.4	65.8
		200–250	262	303.9	82.0
		250–355	355	429.4	129.8
		355–500	486	585.1	154.1
		> 500	744	884.3	252.3
	ball mill		3.8	7.3†	8.7†
	jet mill		1.2	1.5†	7.4†
Flexicoke		< 80			
		80–160			
		160–200			
		200–250	232	234.3	39.8
		250–355	299	302.0	99.5
		355–500			

† Values with limited accuracy due to falling below the detection limit.

The Flexicoke shipment consisted of a powder with small spherical particles having a characteristic particle size in the order of 100 μm. Grinding was therefore not necessary and the coke could be sieved as received using the sieve tower described above. Figure 2.4 shows an SEM picture of Flexicoke particles with an initial particle size of 250-355 μm before (a) and after (b) gasification in the DFB reactor. The unreacted particles are spherical with an onion type of internal structure originating from the layer-wise solid build-up in the Flexicoking reactor. The surface is initially smooth. As carbon conversion increases, the solid is overdrawn with deep cracks perpendicular to the outer particle surface [cf. Figure 2.4 (b)] subdividing the coke into smaller units.

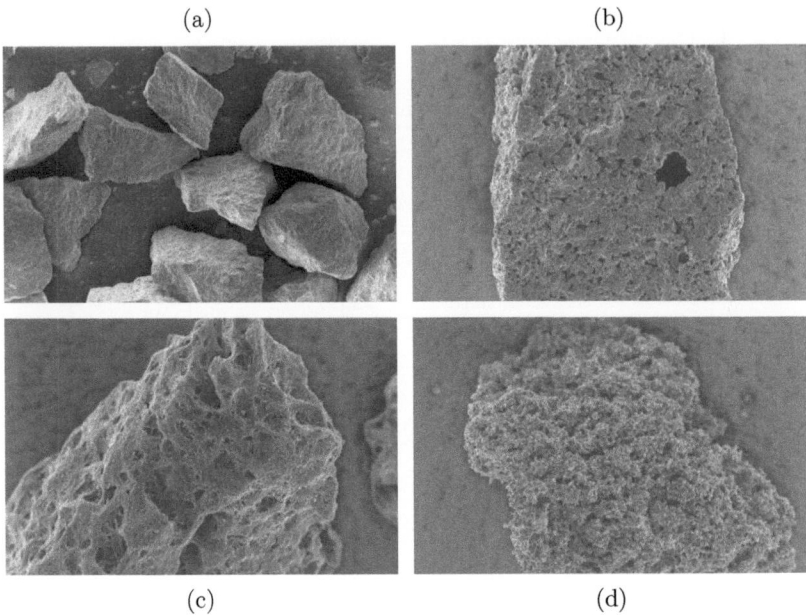

Figure 2.3: SEM micrographs of PD coke particles. (a) shows the raw material, (b) are particles gasified in the DFB reactor with $X_C = 0.21$. (c) and (d) are pictures of PD coke particles partially gasified in the thermobalance with $X_C = 0.63$ and 0.88, respectively. Magnification: (a) 200×, (b)-(d) 1000×.

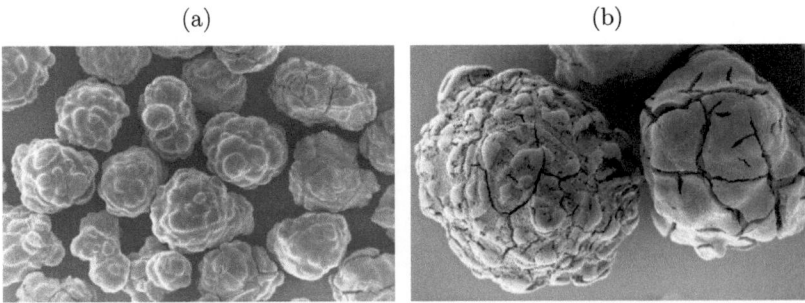

Figure 2.4: SEM micrographs of Flexicoke particles. (a) shows the raw material, (b) are particles gasified in the DFB reactor with $X_C = 0.11$. Magnification: (a) 200×, (b) 500 ×.

Chapter 3

Thermodynamics[1]

This chapter presents the chemical thermodynamics of the steam gasification of petcoke. Chemical equilibrium compositions are computed over a wide range of temperatures. A 2nd-law analysis is conducted to determine the maximum solar exergy conversion efficiency and to identify the major sources of irreversibility.

Together with the reaction kinetics presented in Chapters 4, 5, and 6, this information determines the constrains to be imposed on the design and efficient operation of the solar chemical reactor.

3.1 Equilibrium composition

The steam gasification of petcoke is a complex process, but the overall chemical conversion can be represented by the simplified net reaction:

$$CH_xO_y + (1-y)H_2O = \left(\frac{x}{2} + 1 - y\right)H_2 + CO \qquad (3.1)$$

where x and y are the elemental molar ratios of H/C and O/C in petcoke, respectively. The chemical product is syngas, whose quality depends on x and y. Table 2.2 on page 19 shows the approximate chemical composition, the low heating value, and the elemental molar ratios for Flexicoke and PD coke.

---

[1] Material from this chapter has been published in 'D. Trommer, F. Noembrini, M. Fasciana, et al. Hydrogen production by steam gasification of petroleum coke using concentrated solar power I. Thermodynamic and kinetic analyses. *International Journal of Hydrogen Energy*, 2005. 30(6): p. 605-618', and in 'D. Trommer, M. Romero, H. Rivas, and A. Steinfeld. Hydrogen production via steam gasification of petroleum coke using concentrated solar power. in AchemAmerica. 2003. Mexico City'.

In the analysis that follows, sulfur compounds and other impurities contained in the raw materials are omitted. These are of course important, but it is assumed that their exclusion does not affect the main conclusions. The moisture content (cf. Table 2.1 on page 19) is also not accounted for in Equation (3.1), but the stoichiometric addition of water can be adjusted accordingly.

Equation (3.1) summarizes the overall reaction, but a group of competing intermediate reactions that are essential for the successful gasification need to be considered, namely:

steam gasification:
$$C_{(gr)} + H_2O = CO + H_2 \tag{3.2}$$

Boudouard equilibrium:
$$2\,CO = C_{(gr)} + CO_2 \tag{3.3}$$

methanation:
$$C_{(gr)} + 2\,H_2 = CH_4 \tag{3.4}$$

reforming:
$$CH_4 + H_2O = CO + 3\,H_2 \tag{3.5}$$

water-gas shift:
$$CO + H_2O = CO_2 + H_2 \tag{3.6}$$

all of which depend strongly on the temperature as well as on the pressure and the carbon/oxygen ratio, and determine the relative amounts of H_2, H_2O, CO, CO_2, CH_4, in the gas phase, and $C_{(gr)}$ in the solid phase. The HSC Outokumpu code [72] was used to compute the equilibrium composition of the system $CH_xO_y + (1-y)H_2O$ at 1 bar and over the range of temperatures of interest.

Figure 3.1 and 3.2 show the results for Flexicoke and PD coke, respectively. Species whose mole fraction is less than 10^{-5} have been omitted from the figures. Below about 700 K, CH_4, CO_2, H_2O are the thermodynamically stable components. In the temperature range 800-1100 K, they are used up by a combination of reactions (3.1) to (3.6). When the gasification goes to completion, at above about 1300 K, the chemical system consists of a single gas phase containing H_2 and CO in a molar ratio equal to $(x/2+1-y)$, 1.03 for Flexicoke and 1.17 for PD coke. Figure 3.3 shows the percent yield of H_2 and CO as a function of temperature.

At above 1300 K, the yield for both types of coke exceeds 99%. At higher pressures, as preferred in industrial applications, the thermodynamic

3.1. EQUILIBRIUM COMPOSITION

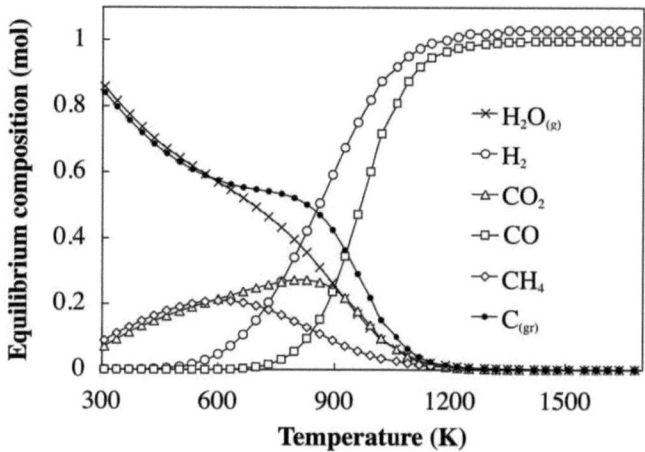

Figure 3.1: Variation of the thermodynamic equilibrium composition with temperature of the system $CH_xO_y + (1-y)H_2O$ at 1 bar for Flexicoke.

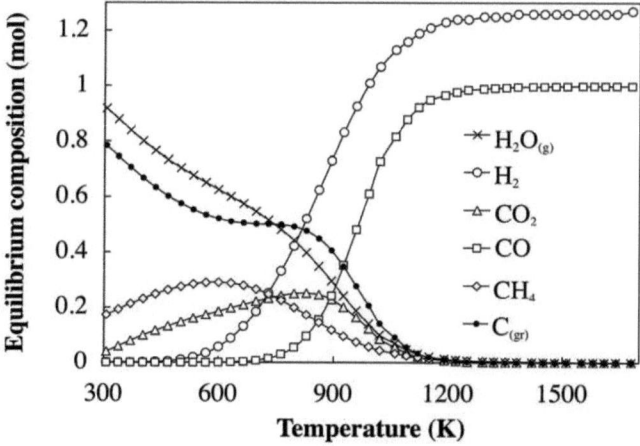

Figure 3.2: Variation of the thermodynamic equilibrium composition with temperature of the system $CH_xO_y + (1-y)H_2O$ at 1 bar for PD coke.

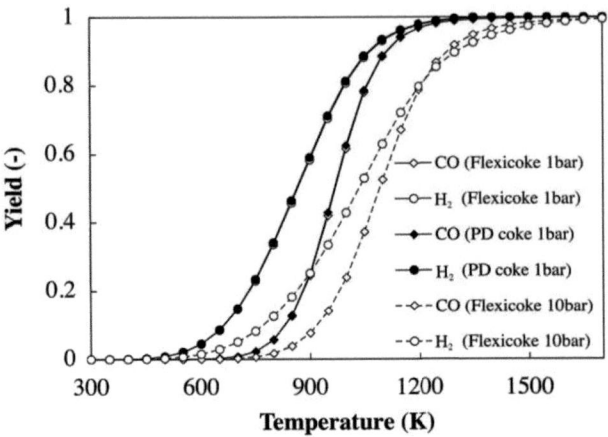

Figure 3.3: Variation of the percent yield of H_2 and CO with temperature for the gasification of Flexicoke and PD coke, assuming the equilibrium composition in Figures 3.1 and 3.2, respectively.

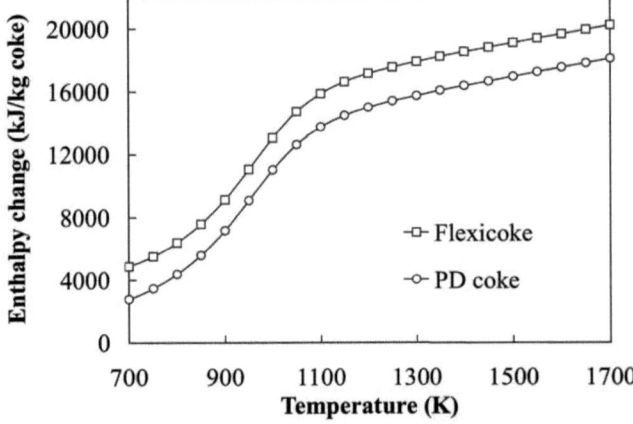

Figure 3.4: Enthalpy change of reaction (3.2) as a function of temperature T for the two types of petcoke, when the reactants are fed at 300 K and the products are obtained at T having the equilibrium composition given in Figures 3.1 and 3.2.

equilibrium of Equation (3.1) is shifted to the left in such a way so as to relieve the pressure in accordance with Le Châtelier's principle. For example, at 10 bar, the equilibrium composition for Flexicoke is shifted such that the gasification goes to 99% completion at above 1600 K, as shown also in Figure 3.3. The results for the PD coke are comparable and the same conclusions can be drawn regarding temperature requirements at higher pressures.

3.2 Enthalpy of reaction

Reaction (3.1) proceeds endothermically in the 800-1500 K range. Figure 3.4 shows the enthalpy change of reaction (3.1) as a function of temperature T for the two types of petcoke, when the reactants are fed at 300 K and the products are obtained at T and having an equilibrium composition given in Figures 3.1 and 3.2, respectively.

Note that since petcoke has no reference enthalpy, all enthalpy changes were calculated by assigning to petcoke (of stoichiometric formula CH_xO_y, with x and y given in Table 2.2) the reference enthalpy of $\{C_{(gr)} + x/2H_2 + y/2O_2\}$ at 300 K, and further adjusting for the small offset between the reported heating value of petcoke and the enthalpy change of reaction (3.7) at 300 K:

$$\underbrace{\left[C_{(gr)} + \frac{x}{2}H_2 + \frac{y}{2}O_2\right]}_{CH_xO_y} + \left(1 + \frac{x}{4} - \frac{y}{2}\right)O_2$$

$$= \frac{x}{2}H_2O_{(g)} + CO_2 \qquad (3.7)$$

At 1300 K, gasifying Flexicoke requires about 18 000 kJ/kg, 2 100 kJ/kg more than the energy required for gasifying PD coke. Part of the reason for the different gasification enthalpies arises from the fact that more water is needed per unit weight of petcoke gasified as the elementary carbon content increases (see Equation (3.1) and Table 2.2), and consequently more energy is needed for producing steam at the required reaction temperature. In summary, the steam gasification process consumes about 56% and 44% of the LHV of the feedstock for Flexicoke and PD coke, respectively.

3.3 Second law analysis

A 2nd-law (exergy) analysis is performed to examine two technically viable routes for extracting power from the chemical products of the gasification, and to establish a base for comparing them with electricity generation by conventional power plants, especially in terms of their CO_2 mitigation potential. The analysis follows the derivation and notation described by v. Zedtwitz and Steinfeld [95].

3.3.1 System setup

The complete process flow sheet is shown in Figure 3.5, with several optional routes grouped in boxes. Box A depicts the solar petcoke gasification process and includes the solar reactor, a heat exchanger, and a quencher. Boxes B_1 and B_2 depict two technically feasible routes for generating electricity from the chemical products of the solar petcoke gasification. These two routes are:

1. (box $A + B_1$) syngas produced by solar steam gasification of petcoke is used to fuel a 55%-efficient combined Brayton-Rankine cycle; and

2. (box $A + B_2$) syngas produced by solar steam-gasification of petcoke is further processed to H_2 by water-gas shift followed by H_2/CO_2 separation, and the H_2 is used to fuel a 65%-efficient fuel cell.

The H_2/CO_2 separation unit is assumed to be based on the pressure swing adsorption technique (PSA) at 90% recovery rate [73]. Its minimum energy expenditure is equal to the ΔG of unmixing, about 1% of the electric output of the fuel cell. State-of-the-art stationary SOFC fuel cells feature energy conversion efficiencies in the range 55-60% when fed with natural gas, and in the range 65-70% when fed directly with hydrogen since the relative loss in the reformer is in the order of 10% [39].

Finally, box C depicts the conventional route for electricity generation by using petcoke as a combustion fuel to a 35%-efficient Rankine cycle.

3.3.2 Results

Table 3.1 shows the complete energy balance and efficiencies calculation. Values for power are normalized to a petcoke mass flow rate of 1 g/s. The three routes considered for generating electricity are depicted by the boxes $A + B_1$, $A + B_2$, and C of Figure 3.5.

The calculations have been carried out using the following baseline parameters for box A: $\dot{m}_{petcoke} = 1$ g/s, $T_{reactor} = 1350$ K, $P_{tot} = 1$ bar, $I = 1$ kW/m², $C = 2\,000$. Further, in box B_1: $\eta_{C.C.} = 55\%$, in box B_2: $\eta_{F.C.} = 65\%$, PSA with 90% recovery rate, and in box C: $\eta_{R.C.} = 35\%$.

3.3. SECOND LAW ANALYSIS

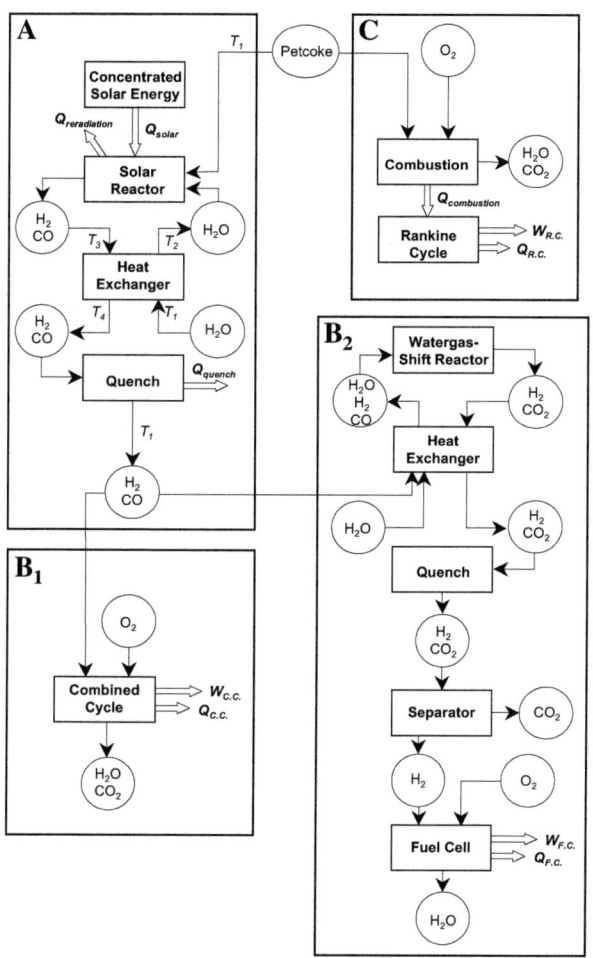

Figure 3.5: Flow sheet diagram used for the 2nd-law analysis. Box A depicts the solar petcoke gasification process. Boxes B_1 and B_2 depict two technically feasible routes for generating electricity from the chemical products of box A, namely: (1) (Box B_1) syngas is used to fuel a combined Brayton-Rankine cycle; and (2) (Box B_2) syngas is processed to H_2 and used to fuel a H_2/O_2 fuel cell. Box C depicts the conventional route of using petcoke to fuel a Rankine cycle.

Table 3.1: Exergy analysis of the solar steam gasification of petcoke using the process modeling shown in Figure 3.5. Three routes for generating electricity are considered: 1) Route A + B$_1$: solar petcoke gasification to syngas, which is used to fuel a combined Brayton-Rankine cycle. 2) Route A + B$_2$: solar petcoke gasification to syngas and further processing to H$_2$, which is used to fuel an H$_2$/O$_2$ fuel cell. 3) Route C: petcoke is used to fuel a Rankine cycle.

	Units	Flexicoke	PD coke
Q_{solar}	kW	12.93	11.02
$Q_{reradiation}$	kW	1.22	1.04
$Q_{reactor,net}$	kW	11.71	9.98
$Q_{heat\ exchanger}$	kW	2.89	2.65
Q_{quench}	kW	2.35	2.72
$W_{C.C.}$	kW$_e$	23.21	23.68
$Q_{C.C.}$	kW	18.99	19.38
$W_{F.C.}$	kW$_e$	21.31	21.92
$Q_{F.C.}$	kW	1.93	1.99
$W_{R.C.}$	kW$_e$	11.54	12.56
$Q_{R.C.}$	kW	21.44	23.32
$\eta_{absorption}$	-	0.906	0.906
$\eta_{heat\ exchanger}$ box A	-	0.552	0.494
$\eta_{heat\ exchanger}$ box B$_2$	-	0.615	0.618
η_{exergy} for route A + B$_1$	-	0.506	0.505
η_{exergy} for route A + B$_2$	-	0.464	0.467
η_{exergy} for route C	-	0.35	0.35
E.O.[a] for route A + B$_1$	kWh$_e$/kg	6.45	6.58
E.O. for route A + B$_2$	kWh$_e$/kg	5.92	6.09
E.O. for route C	kWh$_e$/kg	3.21	3.49
EGF[b] for route A + B$_1$	-	2.01	1.89
EGF for route A + B$_2$	-	1.85	1.75
CO$_2$ emissions A + B$_1$	kg CO$_2$/kWh$_e$	0.54	0.49
CO$_2$ emissions A + B$_2$	kg CO$_2$/kWh$_e$	0.59	0.53
CO$_2$ emissions C	kg CO$_2$/kWh$_e$	1.09	0.93

[a] E.O. = specific electric output.
[b] EGF = electric gain factor; ratio of the electric output of the process to that obtained when using the same amount of petcoke to fuel a 35%-efficient Rankine cycle.

3.3. SECOND LAW ANALYSIS

The heat exchanger of box A is employed for producing steam at 1250 K. Thus, steam and petcoke are fed to the solar reactor at 1250 and 300 K, respectively, and the products of the solar reactor are cooled in the heat exchanger from 1350 K to 792 K and 852 K for Flexicoke and PD coke, respectively, before entering the quencher. The heat exchange of box B_2, with $\eta_{heat\ exchanger} = 62\%$, is employed to preheat the reactants to the water-gas shift reactor from 300 to 600 K.

The maximum exergy efficiency, defined as the ratio of electric power output to the thermal power input (solar power + heating value of reactants), amounts to 51% for route $A + B_1$ and 47% for route $A + B_2$. Thus, both solar gasification routes offer significantly higher exergy efficiencies than the conventional 35%-efficient route via the petcoke-fired Rankine cycle. The path to H_2 and its use in high-efficient fuel cells (route $A + B_2$) is thermodynamically less favorable than using syngas directly in a combined cycle (route $A + B_1$), due the energy consumption in the processing of H_2 from syngas, especially by the H_2/CO_2 separation.

Hydrogen and its use in fuel cells may be the preferable path for mobile and/or decentralized applications. In contrast, syngas may be the desired final fuel for centralized power generation, or for applications in the chemical industry, e.g. synthesis of methanol and Fischer-Tropsch chemicals. Syngas may also be fed directly to SOFC without the need for water-gas shift and gas separation stages, increasing the overall efficiency.

About 9.4% of the input solar power is lost by re-radiation and may be further reduced by increasing the solar concentration ratio. The major source of irreversibility is, however, associated with the quenching of syngas exiting the reactor: 41% and 49% of the solar power input may be lost by quenching the syngas from 1350 to 300 K, for Flexicoke and PD coke respectively. Thus, the use of the heat exchanger of Box A for recovering part of the sensible heat of the syngas becomes justifiable.

Not considered in this analysis is the energy requirement for removing sulfur-containing compounds and other impurities from the syngas, by scrubbing H_2S or by other techniques. In general, the exergy efficiency of a solar energy conversion process is an important criterion for judging its relative economic potential: the higher it is, the lower is the required solar collection area for producing a given amount of fuel, and, consequently, the lower are the costs of the solar concentrating system which usually correspond to about half of the total investments for the entire solar plant [81].

Besides the exergy efficiency, an important figure-of-merit of the process performance is the Electric Gain Factor (EGF), defined as the ratio of the electric output of the process to that obtained when using the same amount of petcoke as a combustion fuel in a 35%-efficient Rankine cycle. The EGF for the two electricity generation routes $A + B_1$ and $A + B_2$ are indicated in Table 3.1 (EGF for route C is obviously 1), along with the specific CO_2 emissions in kg CO_2 per kWh$_e$.

As expected, the EGF is highest for route $A + B_1$ and reaches values of 2.01 and 1.89 for Flexicoke and PD coke, respectively. The EGF is relatively high also for route $A + B_2$, reaching values of 1.85 and 1.75, respectively. Thus, both solar gasification routes offer a remarkable net gain in the electrical output, averaging about double specific electric output vis-à-vis the direct use of petcoke for fueling a 35%-efficient Rankine cycle. Doubling the specific electric output implies reducing by half the CO_2 emissions. Specific CO_2 emissions amount to 0.49 - 0.56 kg CO_2/kWh$_e$, about half as much as the specific emissions discharged by petcoke-fired power plants. Further, for the fuel cell route, if after separating the streams of H_2 and CO_2 the latter is sequestered, the proposed process becomes a decarbonization process that converts petcoke to hydrogen without any release of CO_2 to the atmosphere.

3.4 Summary and conclusions

This chapter presents an analysis of the thermodynamics of the steam-gasification of petcoke using concentrated solar power. The equilibrium composition of the stoichiometric system of petcoke and steam at 1 bar and 1300 K consists of a single gas phase containing H_2 and CO in a molar ratio equal to 1.03 and 1.17 for Flexicoke for PD coke, respectively. A 2nd-law analysis for generating electricity using the gasification products indicates the potential of doubling the specific electrical output and, consequently, halving the specific CO_2 emissions, vis-à-vis conventional petcoke-fired power plants.

In summary, the proposed petcoke/solar hybrid thermochemical processes conserves petroleum and reduces emissions. It further converts intermittent solar energy into a storable and transportable chemical fuel. At the same time, petcoke is solar-upgraded to a cleaner fluid fuel that can be used for electricity generation in highly efficient combined cycles or fuel cells. In contrast to the gasification using process heat derived from the internal petcoke combustion, the solar-driven gasification produces high quality syngas that is not contaminated by the products of combustion. Syngas can be further processed to separate streams of H_2 and CO_2; if the latter is sequestered, the proposed process becomes a decarbonization process that converts petcoke to hydrogen without release of CO_2 to the atmosphere.

Chapter 4

Reaction kinetics[1]

4.1 Heterogeneous gas-solid reactions

The modeling of heterogeneous chemical reactions is complicated by a series of issues not encountered in homogeneous systems. Since more than one phase is present, rate equations for heterogeneous reactions often incorporate mass transfer terms in addition to the usual chemical kinetics term. The mass transfer terms may differ in type and number depending on the shape of the heterogeneous system. A general rate expression that holds for all reacting systems is not available.

For a fluid-particle reaction as it is encountered in the H_2O-CO_2 gasification of coke, the following set of consecutive reaction steps (resistances) can be identified:

Step I Diffusion of the reactant from the bulk gas through the gas film to the external particle surface.

Step II Diffusion of the reactant from the particle surface through the pores to a reaction site.

Step IIIA Reactant adsorbs on the surface and is attached to an active site.

Step IIIB Reaction with the solid or a molecule from the gas phase (single site mechanism) or with an adjacent active site (dual site mechanism).

---

[1]Material from this chapter has been published in 'D. Trommer and A. Steinfeld. Kinetic modeling for the combined pyrolysis and steam gasification of petroleum coke and experimental determination of the rate constants by dynamic thermogravimetry in the 500-1520 K range. *Energy & Fuels*, 2006. 20(3): p. 1250-1258'.

Step IIIC Desorption of the products from the solid surface.

Step IV Diffusion of the products from the reaction site across the pores to the particle surface.

Step V Diffusion of the products from the external particle surface to the bulk gas.

Steps I and V describe the mass transfer of gaseous reactants and products across the fluid film surrounding the coke particle, described in Chapter 4.1.2. The structural particle model in Chapter 4.1.1 considers the steps II and IV. Steps IIIA to IIIC are considered by the Langmuir-Hinshelwood reaction model and treated in Chapter 4.3. The above list could be further extended by an ash layer diffusion resistance, which is neglected since it has no practical relevance for the coke used in this study.

Depending on the physical and chemical boundary conditions of the system such as temperature, pressure, stoichiometry of the reaction, and solid porosity, different resistances from the list above can become rate controlling since it is always the slowest step that controls the overall rate of the process. For the measurement of *intrinsic chemical kinetics* (i.e. based on reaction rates that are not affected by mass transfer limitations), it is desirable to choose the experimental conditions in order to obtain chemical reaction control, which is usually encountered at low temperatures. If this is not feasible diffusion effects have to be considered in the model, leading to undesired complication.

Another important issue is the *contacting pattern* of the involved phases. There are many possibilities to bring two phases in contact with each other. Real chemical reactors can usually be described using the assumption of an idealized flow pattern such as batch, semi-batch, mixed flow or plug flow models. It is essential to have a mathematical model that correctly reproduces the physical situation of the reacting system. A mismatch in the flow model, for example, can easily result in a situation where the reaction kinetics have been simulated with an amazing degree of accuracy and just the same the resulting equations fail to give a correct prediction of a scale-up [49].

In the following chapters the models for the heterogeneous gasification reaction of the H_2O-CO_2-coke system are presented starting on the particle scale with a structural model for a single coke particle (Chapter 4.1.1) and followed by the mass transfer model for the surrounding gas film (Chapter 4.1.2). Chapter 4.2 presents a survey of reactor configurations that are used

4.1. HETEROGENEOUS GAS-SOLID REACTIONS

for the experimental campaigns of this thesis, including the models for the respective flow patterns.

4.1.1 Structural models for the coke particle

There are two important concepts in modeling of non-catalytic gas-solid reactions: The *progressive-conversion* model (PCM) and the *shrinking unreacted core* model (SCM). The PCM and SCM models represent two idealizations that mark the extremes with respect to the reaction behavior of a solid particle and delimit the burn-off characteristics of real particle reacting with a surrounding fluid [49, 99].

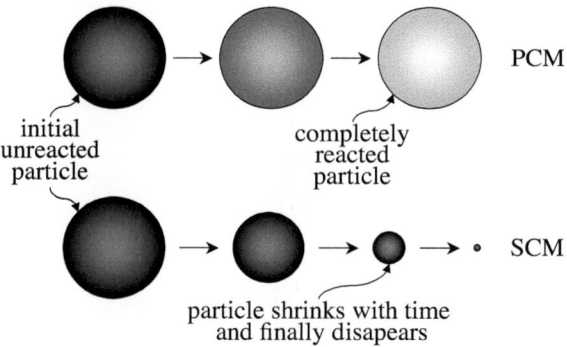

Figure 4.1: Progressive-conversion and shrinking unreacted core model.

Progressive-conversion model. The progressive-conversion model assumes that the diffusion rates in the gas and solid phase are very fast compared to the reaction rate. The concentration of the gaseous reactant inside the particle is everywhere the same and the overall gas-solid reaction rate is solely controlled by the intrinsic reactivity of the solid. While the reaction goes to completion, the solid particle is uniformly converted to solid and/or fluid products, but the particle size rests unaltered as the reaction goes to completion (cf. PCM in Figure 4.1).

Shrinking unreacted core model. The shrinking-core model applies to non-porous particles and situations where the rate of reaction is fast compared to the diffusion rate in the solid. The reaction thereby occurs at the outer surface of the unreacted particle. The reaction zone has the form of a sharp line moving into the solid and leaving behind completely converted as well as inert material (cf. SCM in Figure 4.1). Two cases are possible: (1) No inert material is present, and the solid is converted to a gas without the formation of residues. This situation applies to the steam gasification of petcoke which has a very low ash content and is essentially completely gasified. (2) For materials with a high content of inert material, a solid ash layer is formed inhibiting further conversion due to diffusion limitation inside the particle.

General model

The PCM and SCM models are frequently used idealizations for non-catalytic gas solid reactions applying to situations where either the rate of the gas-solid reaction in the particle is very fast compared to diffusion (SCM), or the diffusion process is very fast compared to the rate of the chemical reaction (PCM). In reality, a porous solid can exhibit the characteristics of both models depending on the particle temperature: At low temperatures, the reaction is slow and the particle is best described by the SCM model. At high temperatures, the reaction rate is usually much faster than diffusion resulting in PCM behavior.

Diffusion with homogeneous reaction. If the reaction and diffusion rates inside a porous particle are of the same order of magnitude, an intermediate case between SCM and PCM arises with the reaction occurring in a diffuse zone, which can be as big as the entire particle. This case is schematically shown in Figure 4.2: A gaseous reactant at the same time diffuses into a spherical particle and reacts with the porous solid causing a characteristic concentration gradient.

The concentration profile across a porous sphere in steady state is the result of the differential mass conservation equation obtained from a mass balance over a spherical shell of thickness ΔR (cf. Figure 4.2) and taking the limit for $\Delta R \to 0$ [11]:

$$\frac{1}{R^2}\frac{d}{dR}\left(R^2\, n_i''\right) = r_i''(p_j, T) \qquad i,j = \mathrm{H_2O},\ \mathrm{H_2},\ \ldots \qquad (4.1)$$

n_i'' is the molar flux of species i per unit surface and time of a sphere with radius R and $r_i''(p_j, T)$ is the homogeneous rate of reaction per unit volume.

4.1. HETEROGENEOUS GAS-SOLID REACTIONS

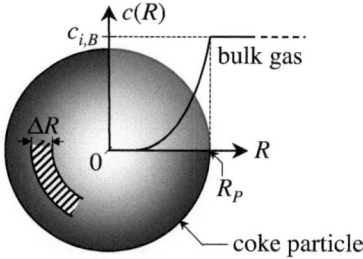

Figure 4.2: Schematic of a spherical particle with concentration gradient and spherical shell of thickness ΔR.

The reaction rate is usually a complicated nonlinear function of the temperature and the partial pressures of reactants and products in the gas phase. The dependence on temperature and partial pressures is defined by the rate laws, which are elaborated in detail in Chapter 4.3.

Assuming that mass transfer inside the porous solid occurs by diffusion only and introducing D_{eff} as the effective diffusivity in the solid matrix [98] yields the diffusion equation according to

$$n_i'' = -D_{\mathit{eff}} \frac{dc_i}{dR} \quad \left(\frac{\mathrm{mol}}{\mathrm{m}^2\,\mathrm{s}}\right) \tag{4.2}$$

where c_i (mol/m^3) is the molar concentration of reactant i inside the pores. Equation (4.2) is inserted in (4.1) to yield the differential transport equation for constant diffusivity,

$$-D_{\mathit{eff}} \frac{1}{R^2} \frac{d}{dR}\left(R^2 \frac{dc_i}{dR}\right) = r_i''(p_j, T) \qquad i, j = \mathrm{H_2O}, \mathrm{H_2}, \ldots \tag{4.3}$$

Equation (4.3) is a standard second order differential equation that is solved with the following boundary conditions:

$$c_i(R = R_P) = \frac{p_{i,B}}{\mathcal{R}\,T} \tag{4.4}$$

$$\left.\frac{dc_i}{dR}\right|_{R=0} = 0 \tag{4.5}$$

Grain model. Szekely and Evans [88, 89] originally developed the grain model for gas-solid reactions with a moving boundary. Later on, the model

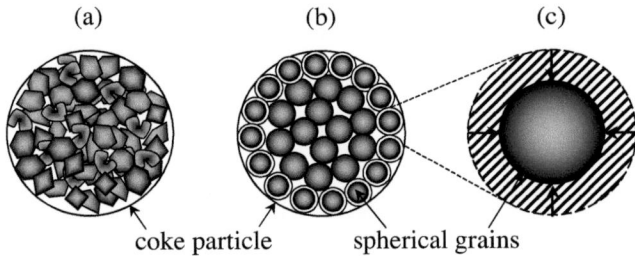

Figure 4.3: Schematic of the grain model. The coke particle is assumed to be an aggregation of smaller grains (a). The grains are modeled as spheres of uniform size (b), each of which reacts with steam according to the SCM model (c).

was successfully applied by Kajitani et al. [40] to describe the rate of the coal char gasification reaction with CO_2 in a pressurized drop tube furnace.

The grain model is used to adapt the aforementioned mathematical model considering diffusion with homogeneous reaction to situations with heterogeneous surface reactions. Heterogeneous reactions often have a fractional reaction order $n < 1$ of the solid rather than $n = 1$ because only part of the solid is accessible to the reaction.

The basic assumption of the grain model is that the solid behaves like an agglomeration of smaller units, the so called *grains*. Figure 4.3 (a) is a schematic of a generic coke particle considered as an aggregation of smaller, randomly shaped grains. The individual grains are modeled as non-porous spheres of uniform size [Figure 4.3 (b)], each of which reacts with the surrounding gas according to the shrinking core model [Figure 4.3 (c)].

Following the definition of the SCM model, the heterogeneous carbon gasification reaction taking place at the external surface of a single grain can be defined as

$$-r_C = -\frac{1}{A_{grain}}\frac{dn_C}{dt} \qquad (4.6)$$

For a spherical grain, both the change of the mole number, dn_i, and the carbon conversion, X_C, of a single *grain* depend on the grain radius as follows

4.1. HETEROGENEOUS GAS-SOLID REACTIONS

$$dn_i = \rho_v dV = \rho_v 4\pi R^2 dR \tag{4.7}$$

$$X_C = 1 - \frac{n_C}{n_{C,0}} = 1 - \left(\frac{R}{R_0}\right)^3 \tag{4.8}$$

Therefore, assuming SCM behavior, the heterogeneous gasification reaction taking place at the external surface of the coke grain becomes

$$-r_C = -\frac{1}{A_{grain}}\frac{dn_C}{dt} = a_{v,0}^{-1}\rho_v(1-X_C)^{-2/3}\frac{dX_C}{dt} \tag{4.9}$$

where $a_{v,0}$ is the surface to volume ratio of the grain at $X = 0$ and ρ_v is the volumetric molar density of carbon.

Equation (4.9) is the surface specific reaction rate of a single grain. It can be used to express the homogeneous reaction inside the coke *particle*, r_i'', as it appears in Equation (4.3), in terms of a pseudo homogeneous surface reaction based on the coke's specific volumetric surface. The specific volumetric surface of the coke, a_v, is defined as

$$a_v = \frac{A_{coke}}{V_{coke}} \tag{4.10}$$

and therefore the homogeneous rate, r_i'', becomes

$$r_i'' = \frac{A_{coke}}{V_{coke}} r_i = a_{v,coke}\, r_i \tag{4.11}$$

$$\text{and for C:} \quad -r_C'' = \rho_v(1-X_C)^{-2/3}\frac{dX_C}{dt} \tag{4.12}$$

Substituting Equation (4.12) into Equation (4.3) yields the governing equation of the grain model

$$-D_{eff}\frac{1}{R^2}\frac{d}{dR}\left(R^2\frac{dc_i}{dR}\right) = a_v r_i \tag{4.13}$$

The solution to differential equations of this type describing diffusion inside the particle in the presence of a chemical reaction is obtained in the form of a radial concentration profile, $c(R)$. Usually, the concentration profile is given in terms of the dimensionless *Thiele Modulus*

$$\phi_M = L\sqrt{\frac{n+1}{2}\frac{k(T)c_i^{n-1}}{D_{eff}(T)}} \tag{4.14}$$

where L is a characteristic spacial dimension of the particle, n is the reaction order, and $k(T)$ a reaction constant.

Since the concentration profile is only of theoretical interest, sources frequently report the effect of mass transfer limitations inside the solid in the form of the particle effectiveness factor, η_P. The particle effectiveness factor is defined as the ratio of the observed overall reaction rate, r_{obs}, and the intrinsic rate of reaction, r_{intr}, that would be observed under the same conditions as r_{obs} if diffusion inside the solid was instantaneous [49, 11, 16]:

$$\eta_P = \frac{r_{obs}}{r_{intr}} \quad (4.15)$$

Straight forward analytical solutions for η_P in terms of ϕ_M are available for the case of simple n-th order power law kinetics, $r_i = k \cdot p_i^n$ with $n \geq 0$, see [49]. For complicated rate expressions the solution of Equation (4.3) is not readily accessible and requires detailed knowledge of the particle porosity and its change with increasing carbon conversion. Models for complex kinetics further require additional stoichiometric and empirical corrections, see [48, 87].

A pore-level calculation required to solve the above problem is beyond the scope of this thesis. η_P is therefore determined experimentally by TG experiments using samples with different particle sizes. Equation (4.15) is used in combination with an empirical correlation between effectiveness factor and particle size to model the variation of the gasification rate with particle size. The correlation for η_P is shown together with the corresponding experimental data in Chapter 5.2.2 on page 68.

4.1.2 Mass transfer model for the gas phase

A mass transfer model for the transport of reactants and products in the gas phase surrounding the coke particle has to be considered in situations where the film mass transfer has a limiting effect on the overall rate of the chemical reaction. Film diffusion resistances are particularly important in reacting systems characterized by high temperatures and big particles. However, film diffusion is negligible in reacting systems characterized by very small particles and/or slow chemical reactions due to low temperature.

Simple theories for interfacial mass transfer usually assume that a stagnant fluid film exists near every surface [16]. If the reactant diffusing across the film is dilute, the convection induced by diffusion is negligible and a simple expression for the steady state flux can be obtained in terms of a mass

4.1. HETEROGENEOUS GAS-SOLID REACTIONS

transfer coefficient, k_{mtf},

$$n_i'' = k_{mtf} \left(p_{i,I} - p_{i,B} \right) \tag{4.16}$$

where $p_{i,I}$ is the partial pressure of component i at the gas-solid interface and $p_{i,B}$ is the partial pressure in the bulk phase. Different expressions for k_{mtf} according to the theories of film, penetration, and surface renewal are available from [16].

If the rate of the chemical reaction is available in a simple form e.g. as a first or second order reaction or a power law expression, it is possible to obtain an overall rate expression for the series of mass transfer and chemical reaction. The overall rate expression depends on the bulk gas composition and is obtained by eliminating the unknown gas concentrations at the particle surface via equating the rate of the chemical reaction with the rate of the mass transfer.

However, the gasification of petcoke considered in this thesis is complicated by two issues that prevent the use of the aforementioned simple approach to derive a single expression taking into account both chemical reaction and mass transfer:

1. Reactants and products in the considered experimental setups are not dilute and the gasification reaction produces a significant amount of gas at the particle-gas interface. It is therefore inevitable to include convection in the mass transfer model.

2. The Langmuir-Hinshelwood rate equations used in this thesis are complicated expressions of the partial pressures of the involved species [cf. Equations (4.53) and (4.83)] and do not allow to eliminate the unknown interface concentration analytically.

As a consequence, stagnant film models can not be used and a more sophisticated model has to be developed, which considers convection in the boundary layer. Since there is no analytical solution for the unknown interface concentration of the gaseous reactants and products, the series of mass transfer and chemical reaction has to be solved numerically.

The mass transfer model presented in the following paragraph is an adaption of a model presented by Welty et al. [98] for the combustion of pulverized coal particles. The model provides a mathematical expression for the mass transfer of component i across a finite concentration boundary layer. The overall rate of reaction is then obtained by equating this expression with rate of the chemical reaction of component i and calculating the root in terms of the unknown interface composition.

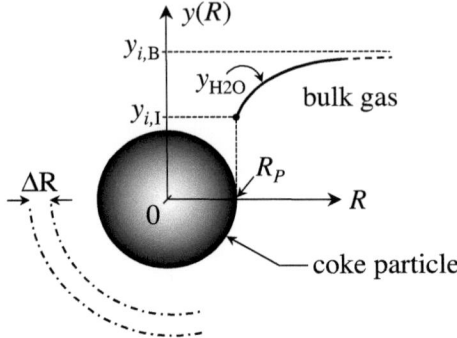

Figure 4.4: Schematic of the mass transfer model for the gas phase: Steam is transfered from the bulk gas to the surface of a spherical petcoke particle, where it reacts with carbon to form H_2, CO, and CO_2.

Film diffusion model

The mass transfer model for the gas phase considers the steady state, one dimensional diffusion and convection of steam near the surface of a petcoke particle, as depicted in the schematic of Figure 4.4. R_P is the radius of the spherical particle and R is the radial coordinate. Steam reacts at the internal and external surfaces of the petcoke particle with solid carbon to form gaseous H_2, CO, and CO_2 according to the following stoichiometric equation:

$$C + (\nu_1 + 2\nu_2) H_2O = \nu_1 CO + \nu_2 CO_2 + (\nu_1 + 2\nu_2) H_2 \qquad (4.17)$$

ν_1 and ν_2 are temperature dependent stoichiometric coefficients (ν_1/ν_2 is the CO/CO_2 molar ratio and $\nu_1 + \nu_2 = 1$ because of the carbon mass balance).

Mass conservation in the spherical film. Because the gas-particle system depicted in Figure 4.4 is in steady state, no mass accumulates in the gas film. Further, taking into account the rotational symmetry of the problem reduces the general continuity equation for species i in spherical coordinates to the following simple differential equation:

$$-\frac{1}{R^2} \frac{d}{dR} \left(R^2 n_i'' \right) = 0 \qquad (4.18)$$

The boundary conditions associated with Equation (4.18) are:

4.1. HETEROGENEOUS GAS-SOLID REACTIONS

- The molar concentration of species i at an infinite distance from the particle surface equals the bulk concentration.

$$y_i(R = \infty) = y_{i,B} = \frac{p_{i,B}}{P_{tot}} \qquad (4.19)$$

- The molar flux of species i at the particle surface, $n_i''(R = R_P)$, equals the overall rate of the gasification reaction throughout the entire coke particle divided by its outer surface.

$$n_i''(R = R_P) = \eta_P \cdot r_i(y_{j,I}, P, T) \frac{m_{coke}\, a}{A_P} \qquad (4.20)$$

The overall rate of the chemical reaction is calculated with Equation (4.15) from the particle model in Chapter 4.1.1 and the rate laws for r_{intr} to be described in Chapter 4.3. The controlling concentration for the chemical reaction is the concentration at the particle-gas interface rather than the bulk concentration as in Chapter 4.1.1.

The molar flux of i at a radial position R, $n_i''(R)$, is obtained from integration of the mass transfer Equation (4.18) and subsequent insertion of the boundary condition (4.20):

$$n_i''(R) = \eta_P \cdot r_i(y_{j,I}, P, T) \frac{m_{coke}\, a}{A_P} \left(\frac{R_P}{R}\right)^2 \qquad (4.21)$$

$$\text{with} \quad [n_i''] = \left(\frac{\text{mol}}{\text{m}^2 \cdot \text{s}}\right) \quad \text{and} \quad i, j = H_2O, H_2, \ldots$$

Flux equation. The molar flux equation for component i in the radial direction is defined as the sum of the diffusive and convective flows [16]

$$n_i'' = j_i'' + y_i \sum_j n_j'' \qquad (4.22)$$

$\sum_j n_j''$ represents the total radial convective flow and j_i'' the molar diffusion flux according to Fick's law

$$j_i'' = -c \cdot D_i \frac{d}{dR} y_i \qquad (4.23)$$

where c is the overall molar concentration of the gas phase.

H_2O is considered as the key transporting species. The respective diffusion Equation (4.22) takes the form

$$n''_{H_2O} = -c \cdot D_{H_2O,m} \frac{dy_{H_2O}}{dR} + y_{H_2O} \left(n''_{H_2O} + n''_{CO} + n''_{CO_2} + n''_{H_2} + n''_{Ar} \right) \quad (4.24)$$

where $D_{H_2O,m}$ is the multicomponent diffusion coefficient of steam into the mixture of Ar, H_2, CO, CO_2 and CH_4. $D_{H_2O,m}$ is calculated using Blanc's law [68] that gives an estimate of the true multicomponent diffusion coefficient being difficult to calculate [16].

$$D_{i,m} = \left(\sum_{\substack{j=1 \\ j \neq i}}^{n} \frac{y_j}{D_{i,j}} \right)^{-1} \quad (4.25)$$

The equations for the fluxes of H_2, CO, and CO_2 result from the stoichiometry of the gasification reaction, Equation (4.17). Argon has no net transfer, since it is an inert [98]. The molar fluxes of the remaining species can be expressed in terms of the steam flux, n''_{H_2O}, as follows:

$$n''_{CO} = -\frac{\nu_1}{\nu_1 + 2\nu_2} n''_{H_2O} \quad (4.26)$$

$$n''_{CO_2} = -\frac{\nu_2}{\nu_1 + 2\nu_2} n''_{H_2O} \quad (4.27)$$

$$n''_{H_2} = -n''_{H_2O} \quad (4.28)$$

Substituting Equations (4.26), (4.27), and (4.28) into (4.24) yields the diffusion equation depending only on the steam flux and the stoichiometric coefficients:

$$n''_{H_2O} = -\frac{c \cdot D_{H_2O,m}}{1 + y_{H_2O}\, \nu_{1,2}} \frac{dy_{H_2O}}{dR} \quad (4.29)$$

with

$$\nu_{1,2} = \frac{\nu_1 + \nu_2}{\nu_1 + 2\nu_2} \quad (4.30)$$

Integration of Equation (4.29) over the gas phase ($R = R_P \to \infty$) results in the flux equation for steam

$$n''_{H_2O}(R) = \frac{R_P}{R^2} \frac{c \cdot D_{H_2O,m}}{\nu_{1,2}} \ln \left(\frac{1 + y_{H_2O,I}\, \nu_{1,2}}{1 + y_{H_2O,B}\, \nu_{1,2}} \right) \quad (4.31)$$

The remaining species, i.e. H_2, CO, and CO_2, are obtained from Equations (4.26) to (4.28).

4.2. EXPERIMENTAL SYSTEMS FOR KINETIC ANALYSES

The amount of steam transfered from the particle to the gas phase is the product of the steam flux and the external surface area of the particle, $A_P = 4\pi R_P^2$:

$$A_P \cdot n''_{H_2O}(R = R_P) = 4\pi R_P \frac{c \cdot D_{H_2O,m}}{\nu_{1,2}} \ln\left(\frac{1 + y_{H_2O,I}\, \nu_{1,2}}{1 + y_{H_2O,B}\, \nu_{1,2}}\right) \quad (4.32)$$

The overall rate of reaction, $r_{i,obs}$, is the result of the series process consisting of the two resistances 'film diffusion' and 'surface reaction' and is obtained from equating Equation (4.32) for the amount of gas transfered in the fluid film with the rate of the gasification reaction elaborated in the previous section:

$$A_P \cdot n''_i(R_P, y_{j,B}, y_{j,I}) = \eta_P\, m_P\, a\, r_i(y_{j,I}, P, T) \quad (4.33)$$

$$\text{with} \quad i, j = H_2O, H_2, \ldots$$

Equation (4.33) is a system of four equations that is solved numerically for the unknown interface composition, $y_{j,I}$. The result is used to calculate the overall rate of reaction with the right hand side of Equation (4.33). A discussion of the effect of the gas phase mass transfer resistance on the overall rate of the gasification reaction is presented in Chapter 8.4.2 for the model of the 5 kW process reactor.

4.2 Experimental systems for kinetic analyses

4.2.1 Differential and integral reactors

The experimental strategy in studying the rate of heterogeneous gas-solid reactions and rate law exploration usually involves measuring the extent of conversion in a semi-batch setup operated with a steady fluid flow. Reactors with any flow pattern (mixed flow, plug flow, etc) can be used, as long as the pattern selected is known; otherwise the kinetics can not be found [49].

Two reactor concepts for the analysis of experimental rate data can be distinguished. They are briefly discussed at this point, as there is a certain relevance to the evaluation of the experimental rate data in the following chapters:

Differential reactor. The basic assumption behind a differential flow reactor is that the reaction rate is constant at all points within the reactor. Since heterogeneous reactions are concentration-dependent, this assumption is only reasonable in the case of small conversion and/or small reactors. An example of a differential reactor is the thermogravimeter discussed in Chapter 5: The samples used for TG experiments are small ($\approx$ 30-40 mg) and mass transfer in the gas phase surrounding the coke particles is considered to be fast. Therefore, no concentration gradients emerge, and the mass conservation equation can be applied in the differential form.

Integral reactor. Integral reactors have a high spacial variation of the reaction rate due to significant chemical conversion, and the fluid passing through the reactor shows significant concentration gradients. Concentration changes in the reactor have to be considered in the method of analysis using the integrated form of the mass conservation equation. Examples of integral reactors are the fluidized bed reactors presented in Chapter 6 and the SynPet reactor discussed in Chapters 7 and 8.

4.2.2 Thermogravimetry

Thermogravimetry (TG, TGA) is a common application used in chemistry and material science for the determination of thermochemical properties of substances and mixtures of substances. In addition, it provides an excellent means to measure the reaction kinetics of various physical and chemical processes involving solids and liquids. Thermogravimetry on its own determines the temperature and time resolved changes of the sample mass while the sample is exposed to a controlled gas atmosphere and a predefined temperature program [33].

TG experiments are performed in a thermobalance, which essentially consists of a high temperature furnace and an electronic precision balance. The furnace chamber is vacuum proof and allows to perform experiments with controlled sample environments that can be reactive or inert. The balance is directly connected with the sample crucible in the furnace and measures the sample weight online as a function of temperature and time. The temperature control of the furnace allows two modes of operation being either isothermal, dynamic with a constant heating rate, or any combination of the two.

The sample arrangement in the thermobalance is of particular importance. If a cup containing a bed of particles with a stagnant gas atmosphere

4.3. REACTION MECHANISMS

on top is used (cf. [65, 28]), the sample has to be considered as an integral reactor and mass and heat transfer equations need to be solved in order to obtain meaningful rate data. If, on the other hand, the sample size is small and mass transfer is fast compared to the chemical reaction as it is assumed for the experiments in Chapter 5, concentration and temperature gradients in the sample and in the surrounding gas phase are negligible allowing to use the differential reactor concept.

Often TG is combined with other analyses: The evolved gas analysis (EGA) using gas and mass chromatography provides additional information on chemical reactions based on the production or consumption of gaseous species. Other thermal analysis techniques such as differential thermal analysis (DTA) and differential scanning calorimetry (DSC) provide data on temperature differences and heat fluxes of the sample and are usually applied in cases where calorific data is required.

4.3 Reaction mechanisms

4.3.1 Pyrolysis

The pyrolysis model used in this thesis, the so called *distribution of activation energy model*, was originally developed by Pitt [67] and later on used in other experimental studies on coal devolatilization [5] and rapid pyrolysis of coal blends [92].

The basic assumptions of the model are:

- Coke is a mixture of pseudo-components decomposing independently.
- The decomposition reaction of the pseudo-components is of first order.
- There exists a wide distribution of the activation energy, and a specific value for the activation energy is associated with a characteristic decomposition temperature.

Based on the above assumptions, the rate of pyrolysis is calculated as a linear combination of first-order decomposition rates of pseudo-components:

$$\frac{dX_P}{dt} = \sum_i x_i r_i = \sum_i x_i \frac{d\alpha_i}{dt} \qquad (4.34)$$

where x_i is the ultimate yield obtained by complete decomposition of component i and the conversion of the individual pseudo-components, r_i, is defined as

$$r_i = \frac{d\alpha_i}{dT}\beta = k_i\left(1 - \alpha_i\right) \qquad (4.35)$$

where the temperature dependence of the rate constant is given by the Arrhenius law,

$$k_i = k_{0,i} \cdot \exp\left(\frac{-E_{A,i}}{\mathcal{R}\,T}\right) \qquad (4.36)$$

k_0 is the pre-exponential factor and E_A is the activation energy in (J/mol).

4.3.2 H_2O-CO_2 gasification

Two mechanisms are considered for the gasification of carbon with H_2O and CO_2: the *oxygen-exchange* and the *extended* mechanisms.

Oxygen exchange mechanism

The oxygen-exchange mechanism postulates reversible oxygen-exchange surface reactions

$$CO_2 + C\langle*\rangle \underset{k'_{-1}}{\overset{k'_1}{\rightleftarrows}} CO + C\langle O\rangle \qquad (4.37)$$

$$H_2O + C\langle*\rangle \underset{k'_{-2}}{\overset{k'_2}{\rightleftarrows}} H_2 + C\langle O\rangle \qquad (4.38)$$

followed by the unidirectional gasification step at higher temperatures through combination of oxygen and carbon atoms from the solid to form $CO(g)$,

$$C\langle O\rangle \xrightarrow{k'_3} CO + nC\langle*\rangle \qquad (4.39)$$

In the above equations, $C\langle*\rangle$ and $C\langle O\rangle$ represent free active sites on the carbon surface and active sites with adsorbed oxygen, respectively.

Inhibition occurs by the recombination of adsorbed oxygen with either CO or H_2. Equation (4.39) typically controls the movement towards equilibrium, while Equations (4.37) and (4.38) provide for the water-gas shift equilibrium. Assuming elementary reactions, the rate laws for the formation/consumption of the gas species H_2O, CO_2, CO, and H_2 are obtained as functions of their partial pressures in the bulk phase and the fraction of surface covered with

4.3. REACTION MECHANISMS

adsorbed oxygen, θ_O:

$$r_{H_2O} = k'_{-2} p_{H_2} \theta_O - k'_2 p_{H_2O}(1-\theta_O) \qquad (4.40)$$
$$r_{CO_2} = k'_{-1} p_{CO} \theta_O - k'_1 p_{CO_2}(1-\theta_O) \qquad (4.41)$$
$$r_{H_2} = k'_2 p_{H_2O}(1-\theta_O) - k'_{-2} p_{H_2} \theta_O \qquad (4.42)$$
$$r_{CO} = k'_1 p_{CO_2}(1-\theta_O) - \left(k'_{-1} p_{CO} - k'_3\right)\theta_O \qquad (4.43)$$

where

$$r_i = \frac{1}{A}\frac{dn_i}{dt} = \frac{1}{m_{coke} \cdot a}\frac{dn_i}{dt} \qquad (4.44)$$

r_i, with $i = $ H$_2$O, CO$_2$, CO, and H$_2$, is expressed in units of moles of i transformed per time per solid surface, (mol/m^2/s). Further assuming sorption equilibrium, i.e.

$$\frac{d\theta_O}{dt} = -r_{H_2O} - r_{CO_2} - k'_3 \theta_O = 0 \qquad (4.45)$$

yields:

$$\theta_O = \frac{k'_1 p_{CO_2} + k'_2 p_{H_2O}}{k'_1 p_{CO_2} + k'_{-1} p_{CO} + k'_2 p_{H_2O} + k'_{-2} p_{H_2} + k'_3} \qquad (4.46)$$

Substituting θ_O in Equations (4.40) to (4.43) gives the expressions of the rate laws as functions of partial pressures and temperature,

$$r_i = \frac{N_i}{1 + \frac{1}{k'_3}\left(k'_1 p_{CO_2} + k'_{-1} p_{CO} + k'_2 p_{H_2O} + k'_{-2} p_{H_2}\right)} \qquad (4.47)$$

with the species dependent numerators, N_i, being

$$N_{H_2O} = -k'_2 p_{H_2O} - \frac{k'_2 k'_{-1}}{k'_3} p_{H_2O} p_{CO} + \frac{k'_1 k'_{-2}}{k'_3} p_{H_2} p_{CO_2} \qquad (4.48)$$
$$N_{H_2} = -N_{H_2O} \qquad (4.49)$$
$$N_{CO_2} = -k'_1 p_{CO_2} + \frac{k'_2 k'_{-1}}{k'_3} p_{H_2O} p_{CO} - \frac{k'_1 k'_{-2}}{k'_3} p_{H_2} p_{CO_2} \qquad (4.50)$$
$$N_{CO} = 2k'_1 p_{CO_2} + k'_2 p_{H_2O} - \frac{k'_2 k'_{-1}}{k'_3} p_{H_2O} p_{CO} + \frac{k'_1 k'_{-2}}{k'_3} p_{H_2} p_{CO_2} \qquad (4.51)$$

The rate for the carbon consumption is obtained by applying carbon mass conservation,

$$\begin{aligned} -r_C &= r_{CO} + r_{CO_2} \\ &= k'_3 \theta_O \\ &= \frac{k'_1 p_{CO_2} + k'_2 p_{H_2O}}{1 + \frac{1}{k'_3}\left(k'_1 p_{CO_2} + k'_{-1} p_{CO} + k'_2 p_{H_2O} + k'_{-2} p_{H_2}\right)} \end{aligned} \qquad (4.52)$$

Since the gaseous products CO and H_2 are continuously withdrawn from the reaction site, the reverse reactions of Equations (4.37) and (4.38) are neglected ($k'_{-1} = k'_{-2} = 0$), leading to:

$$-r_C = \frac{k'_1 p_{CO_2} + k'_2 p_{H_2O}}{1 + \frac{1}{k'_3}(k'_1 p_{CO_2} + k'_2 p_{H_2O})} \qquad (4.53)$$

Extended mechanism

The extended mechanism assumes that H_2O undergoes dissociation at the carbon surface into hydroxyl radical and hydrogen atom groups, both being rapidly chemisorbed at adjacent active sites, followed by a H-transfer to the adsorbed hydrogen [50, 13, 60, 62], and that CO_2 reacts by oxygen-exchange with the solid surface according to the previous mechanism. Inhibition occurs by recombination of CO with adsorbed oxygen and by H_2/CO adsorption on active sites. Considering reversible sorption of gaseous species onto the carbon surface and irreversible reactions among adsorbed species and with molecules in the gas phase:

$$CO_2 + C\langle * \rangle \underset{k_{-1}}{\overset{k_1}{\rightleftarrows}} CO + C\langle O \rangle \qquad (4.54)$$

$$CO + C\langle * \rangle \underset{k_{-2}}{\overset{k_2}{\rightleftarrows}} C\langle CO \rangle \qquad (4.55)$$

$$C\langle O \rangle \xrightarrow{k_3} CO + nC\langle * \rangle \qquad (4.56)$$

$$H_2O + C\langle * \rangle \underset{k_{-4}}{\overset{k_4}{\rightleftarrows}} C\langle OH \rangle + C\langle H \rangle \qquad (4.57)$$

$$C\langle OH \rangle + C\langle H \rangle \xrightarrow{k_5} C\langle O \rangle + C\langle H_2 \rangle \qquad (4.58)$$

$$H_2 + C\langle * \rangle \underset{k_{-6}}{\overset{k_6}{\rightleftarrows}} C\langle H_2 \rangle \qquad (4.59)$$

Equations (4.54) to (4.56) are pertinent to CO_2-gasification, while Equations (4.56) to (4.59) are pertinent to H_2O-gasification, with Equation (4.56) describing desorption of the surface oxygen complex common to both H_2O and CO_2 gasification. Gadsby [26] proposed the system of Equations (4.54 forward), (4.55), and (4.56), in which CO inhibition occurs due to adsorption of CO on the carbon surface, competing with CO_2 for active sites. Ergun and coworkers [21, 57] proposed the system of Equations (4.54) and (4.56), in which CO inhibition occurs by detaching of surface oxygen by CO from the gas phase [Equation (4.54) reverse]. Both mechanisms lead to the same

4.3. REACTION MECHANISMS

form of the rate law for carbon consumption by gasification using only CO_2 as gasifying agent:

$$-r_C = \frac{a_1 p_{CO_2}}{1 + a_2 p_{CO_2} + a_3 p_{CO}} \quad (4.60)$$

with a_1, a_2, and a_3 being lumped rate constants. Reif [70] analyzed in detail the importance of the two CO inhibition paths described above and found experimentally that the recombination of $C\langle O\rangle$ and CO is considerably faster than CO adsorption. For gasification at elevated pressures, Blackwood and Ingeme [12] and Müller [63] extended the Erguns mechanism by adding:

$$CO_2 + C\langle CO\rangle \xrightarrow{k_7} 2CO + C\langle O\rangle \quad (4.61)$$

$$CO + C\langle CO\rangle \xrightarrow{k_8} CO_2 + 2C\langle *\rangle \quad (4.62)$$

In the mathematical treatment that follows, Equations (4.61) and (4.62) are omitted from consideration since the reduced system represents reasonably well data in the 0-1 bar pressure range used in this study. CO inhibition can occur both by recombination with an adsorbed oxygen [Equation (4.54) reverse] and by competition for active sites [Equation (4.55)]. Further assuming sorption equilibrium for CO, O, H_2, and the intermediate component OH-H representing the adsorbed water,

$$\theta_{OH,H}: \qquad k_4 p_{H_2O}\theta_v = (k_{-4} + k_5)\theta_{OH,H} \quad (4.63)$$
$$\theta_{CO}: \qquad k_2 p_{CO}\theta_v = k_{-2}\theta_{CO} \quad (4.64)$$
$$\theta_{H_2}: \qquad k_6 p_{H_2}\theta_v + k_5\theta_{OH,H} = k_{-6}\theta_{H_2} \quad (4.65)$$
$$\theta_O: \qquad k_5\theta_{OH,H} + k_1 p_{CO_2}\theta_v = (k_{-1}p_{CO} + k_3)\theta_O \quad (4.66)$$

with

$$\theta_v = 1 - (\theta_{OH,H} + \theta_{CO} + \theta_{H_2} + \theta_O) \quad (4.67)$$

Assuming elementary reactions, the rate laws for the formation/consumption the gas species H_2O, H_2, CO_2, and CO are obtained as functions of their partial pressures and θ_i:

$$r_{H_2O} = k_{-4}\theta_{OH,H} - k_4 p_{H_2O}\theta_v \quad (4.68)$$
$$r_{H_2} = k_{-6}\theta_{H_2} - k_6 p_{H_2}\theta_v \quad (4.69)$$
$$r_{CO} = k_1 p_{CO_2}\theta_v - k_{-1}p_{CO}\theta_O + k_{-2}\theta_{CO} - k_2 p_{CO}\theta_v + k_3\theta_O \quad (4.70)$$
$$r_{CO_2} = k_{-1}p_{CO}\theta_O - k_1 p_{CO_2}\theta_v \quad (4.71)$$

The system of Equations (4.63) to (4.67) is solved analytically for the unknown θ_i, and the solution inserted in Equations (4.68) to (4.71) to yield the rate expressions of Langmuir-Hinshelwood type:

$$r_i = N_i \left[1 + \frac{k_1}{k_3} p_{CO_2} + \frac{k_4 \left(k_5 k_{-6} + k_3 k_{-6} + k_3 k_5 \right)}{k_3 k_{-6} \left(k_{-4} + k_5 \right)} p_{H_2O} \right.$$
$$+ \frac{k_6}{k_{-6}} p_{H_2} + \left(\frac{k_2}{k_{-2}} + \frac{k_{-1}}{k_3} \right) p_{CO} + \frac{k_{-1} k_2}{k_{-2} k_3} p_{CO}^2 \quad (4.72)$$
$$\left. + \frac{k_{-1} k_4}{k_3 k_{-6}} \frac{k_5 + k_{-6}}{k_{-4} + k_5} p_{H_2O} p_{CO} + \frac{k_{-1} k_6}{k_3 k_{-6}} p_{H_2} p_{CO} \right]^{-1}$$

with the following numerator terms:

$$N_{H_2O} = -\frac{k_4 k_5}{k_{-4} + k_5} p_{H_2O} - \frac{k_{-1}}{k_3} \frac{k_4 k_5}{k_{-4} + k_5} p_{H_2O} p_{CO} \quad (4.73)$$
$$N_{H_2} = -N_{H_2O} \quad (4.74)$$
$$N_{CO} = 2 k_1 p_{CO_2} + \frac{k_4 k_5}{k_{-4} + k_5} p_{H_2O} - \frac{k_{-1}}{k_3} \frac{k_4 k_5}{k_{-4} + k_5} p_{H_2O} p_{CO} \quad (4.75)$$
$$N_{CO_2} = -k_1 p_{CO_2} + \frac{k_{-1}}{k_3} \frac{k_4 k_5}{k_{-4} + k_5} p_{H_2O} p_{CO} \quad (4.76)$$

Equation (4.72) represents the complete solution of the reactions described by the system of Equations (4.54) to (4.59). Neglecting the mixed and squared terms in the denominator since they have negligible effect in the low-pressure range, and replacing the elementary rate constants by lumped complex constants, yields:

$$r_{H_2O} = \frac{-K_2 p_{H_2O} - K_2 K_3 p_{H_2O} p_{CO}}{1 + K_4 p_{CO_2} + K_5 p_{H_2O} + K_6 p_{H_2} + K_7 p_{CO}} \quad (4.77)$$
$$r_{H_2} = -r_{H_2O} \quad (4.78)$$
$$r_{CO} = \frac{2 K_1 p_{CO_2} + K_2 p_{H_2O} - K_2 K_3 p_{H_2O} p_{CO}}{1 + K_4 p_{CO_2} + K_5 p_{H_2O} + K_6 p_{H_2} + K_7 p_{CO}} \quad (4.79)$$
$$r_{CO_2} = \frac{-K_1 p_{CO_2} + K_2 K_3 p_{H_2O} p_{CO}}{1 + K_4 p_{CO_2} + K_5 p_{H_2O} + K_6 p_{H_2} + K_7 p_{CO}} \quad (4.80)$$

4.3. REACTION MECHANISMS

where the K_i are defined as:

$$\begin{aligned} K_1 &= k_1 \\ K_2 &= \frac{k_4 k_5}{k_{-4} + k_5} \\ K_3 &= \frac{k_{-1}}{k_3} \\ K_4 &= \frac{k_1}{k_3} \\ K_5 &= \frac{k_4 (k_5 k_{-6} + k_3 k_{-6} + k_3 k_5)}{k_3 k_{-6} (k_{-4} + k_5)} \\ K_6 &= \frac{k_6}{k_{-6}} \\ K_7 &= \frac{k_2}{k_{-2}} + \frac{k_{-1}}{k_3} \end{aligned} \quad (4.81)$$

The rate of carbon consumption is calculated in analogy to the oxygen-exchange mechanism by applying carbon mass conservation:

$$\begin{aligned} -r_\mathrm{C} &= r_\mathrm{CO} + r_{\mathrm{CO}_2} \\ &= k_3 \theta_\mathrm{O} \\ &= \frac{K_1 p_{\mathrm{CO}_2} + K_2 p_{\mathrm{H_2O}}}{1 + K_4 p_{\mathrm{CO}_2} + K_5 p_{\mathrm{H_2O}} + K_6 p_{\mathrm{H}_2} + K_7 p_\mathrm{CO}} \end{aligned} \quad (4.82)$$

Further neglecting the inhibition effect of the product gases in Equations (4.77) to (4.80) and (4.82), the final expression for the rate of carbon conversion becomes:

$$-r_\mathrm{C} = \frac{K_1 p_{\mathrm{CO}_2} + K_2 p_{\mathrm{H_2O}}}{1 + K_4 p_{\mathrm{CO}_2} + K_5 p_{\mathrm{H_2O}}} \quad (4.83)$$

Note that Equation (4.53) obtained by applying the oxygen-transfer mechanism is a special case of Equation (4.83), with the additional boundary conditions

$$K_4 = \frac{K_1}{a_4} \quad \text{and} \quad K_5 = \frac{K_2}{a_4} \quad (4.84)$$

reducing the number of independent parameters from 4 in Equation (4.83) to only 3 in Equation (4.53). Note also that Equation (4.72) does not consider the dual-site mechanism or the two-center adsorption [97]. The temperature dependence of the rate constants derived by the oxygen-transfer mechanism [Equation (4.53)] and by the extended mechanism [Equation (4.83)] is given by the Arrhenius law [Equation (4.36)].

The experimental determination of the constants K_1, K_2, K_3, K_4, and K_5 in Equations (4.77) to (4.80) and (4.83) by dynamic experiments with a thermobalance is described in Chapter 5. Further, K_1, K_2, and K_3 are determined in Chapter 6 using a fluidized bed reactor heated by direct and indirect thermal radiation from a solar simulator. Using two different modes of heat transfer allows to test if the selection of a direct or indirect irradiated solar reactor system has an effect on the rate of the chemical reaction. Such an effect would cause differences with respect to the measured rate constants.

Chapter 5

Experimental rate data from thermogravimetry[1]

5.1 Experimental

5.1.1 Setup: Netzsch STA 409 and Varian Micro GC

The experiments in this chapter were performed with a thermogravimeter system (Netzsch STA 409 CD) equipped with two furnaces. The first is a conventional high-temperature electric furnace with a maximum working temperature of 1823 K. This furnace is suitable for experiments with controlled reactive gas atmospheres having a dew point below room temperature. The second is a special electric furnace with a maximum working temperature of 1523 K, suitable for reactive atmospheres containing up to 100% steam at 1 bar total pressure. A schematic of the steam-furnace (a) and the high temperature furnace (b) with the main components indicated is shown in Figure 5.1. Both furnaces are mounted on a double hoist system allowing alternative use and offering maximum flexibility with respect to experimental conditions such as temperature and steam concentration.

The gasification experiments were performed with dish shaped sample crucibles, shown in Figure 5.2, that are made of aluminum oxide (Al_2O_3) and have an outer diameter of 17 mm. The sample carrier connecting the crucible with the balance is equipped with a thermocouple type S and measures the

---

[1] Material from this chapter has been published in 'D. Trommer and A. Steinfeld. Kinetic modeling for the combined pyrolysis and steam gasification of petroleum coke and experimental determination of the rate constants by dynamic thermogravimetry in the 500-1520 K range. *Energy & Fuels*, 2006. 20(3): p. 1250-1258'.

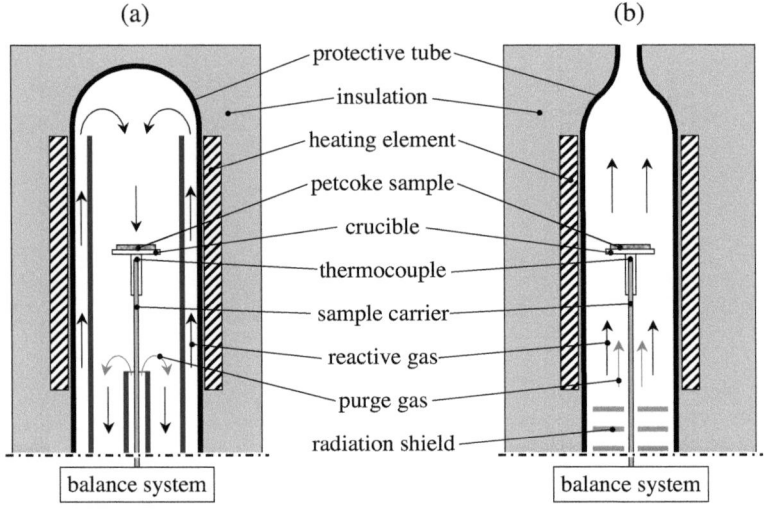

Figure 5.1: Cross-sectional schematics of the Netzsch steam furnace (a) and high temperature furnace (b).

Figure 5.2: Empty crucible (left) and crucible with mounted coke sample (right).

temperature directly at the sample crucible. Petcoke samples of 35 to 40 mg were mounted as a thin layer on the flat sample crucibles and swept by the reacting gas flowing through the furnace. This arrangement avoids stagnant gas around the sample and minimizes mass and heat transfer resistances between the coke sample and the bulk gas [65, 28].

The measuring part of the STA 409 thermogravimeter features a vacuum proof housing suitable to realize controlled atmospheres that can be inert or reactive. The experimental setup includes a TASC 414 System Controller for temperature programming and data acquisition. The weight signal from the

5.1. EXPERIMENTAL

Table 5.1: Analyzed gas mixtures used for the GC calibration

name	manufacturer	nominal composition		balance
		gas	conc.	
	PanGas	$H_2, O_2, N_2,$ CO, CH_4, CO_2	1%	Ar
Mix 216 Scotty 4	Scott Speciality Gases	$CO_2, CO, CH_4,$ C_2H_6, C_2H_4, C_2H_2	1%	N_2
Mix 234 Scotty 14	Scott Speciality Gases	H_2, CH_4 CO_2, CO, N_2, O_2	4% 5%	He
	Praxair	H_2S, COS	1000 ppm	Ar

digital balance and the temperature signal from the sample thermocouple are recorded with a nominal accuracy of $5 \cdot 10^{-3}$ mg and 0.1 K, respectively.

The steam furnace (Figure 5.1 a) is used for the gasification runs with H_2O-CO_2-Ar mixtures. The reactive gas enters the furnace chamber at the top end and flows downwards past the coke sample. The mixing of reactive gas (steam) and purge gas from the balance takes place near the exit of the furnace, offering the possibility to perform experiments with arbitrary steam concentrations up to 100%.

The steam furnace is coupled with a steam generator unit (Bronkhorst Hitec CEM) via a transfer line heated to 200 °C to avoid the condensation of steam. Mechanical flow controllers for Ar and CO_2 (Vögtlin Q-FLOW) and electronic flow controllers for water (Bronkhorst LIQUI-FLOW) are used to set the mass flow rates and the steam concentration in the reactive gas.

The high-temperature furnace (Figure 5.1 b) was used for the pyrolysis runs in pure Ar. In opposition to the steam furnace, this setup has no double furnace wall and features a bottom up flow. The furnace chamber is separated from the balance by means of a ceramic radiation shield, and the gas exit is located at the top end. If a reactive gas is used, the concentration of the reactant is reduced due to the dilution with purge gas coming from the balance. Both the steam and the high-temperature furnaces are suitable to run experiments with corrosive gases as far as material and safety constraints are obeyed.

Every experimental run with the TG setup included an evolved gas analysis in order to retrieve further information about the chemical reactions occurring in the sample. The thermobalance is therefore coupled to a Varian CP

4900 Quad micro gas chromatograph (GC) equipped with a Molecular Sieve 5A PLOT and a PoraPLOT U column (both 10 m with heated injectors). The first column is used to analyze the permanent gases H_2, O_2, N_2, CH_4 and CO using argon as carrier gas. The second is operated with helium as carrier gas and measures $CO_2, C_2H_6, C_2H_4, C_2H_2, H_2S$ and COS. To obtain quantitative concentration data, the micro GC was calibrated in advance using a set of four analyzed calibration gases. The compositions of the respective gas mixtures are indicated in Table 5.1.

Rate data for the H_2O-CO_2 reactive gasification and the pyrolysis reaction is obtained from dynamic TG experiments with constant heating rates. The experimental procedure includes the measurement of the weight loss curve of coke samples exposed to different reactive and inert gas atmospheres [61] and is elaborated in the Chapter 5.1.2.

5.1.2 Experimental procedure

Pyrolysis

The pyrolysis experiments were conducted in the high temperature furnace (cf. Figure 5.1) allowing the investigation of the thermal decomposition over the whole temperature range from ambient up to 1550 K. A non-isothermal (dynamic) temperature program was used starting at 500 K and ramping up with a linear heating rate, $\beta = 10$ Kmin^{-1}, to the final pyrolysis temperature of 1550 K. Experiments are performed with inert sample atmospheres consisting of 100% argon at a flow rate of 210 ml$_N$/min and a total pressure of 1 bar.

An experimental run with the TG setup starts with the activation of the auxiliary devices and loading of the temperature program. A ≈ 36 mg coke sample, weighed on an external Mettler Toledo XS105 DualRange high precision balance is placed as a thin layer on the flat sample crucible, as shown in Figure 5.2, and stored in the TG furnace. The furnace chamber is purged three times with argon to ensure an inert sample atmosphere at the beginning of the experiment. Once this is done, the temperature program is started and automatically completed by the control system.

Table 5.2 presents a list with the pyrolysis runs and the respective temperature and concentration data. Compared to the gasification experiments that are elaborated in the following subsection, samples with smaller particles (200-250 μm) are used for experiments at high temperatures, because bigger samples caused systematic errors due to bursting of single particles.

5.1. EXPERIMENTAL

H_2O-CO_2 gasification

The experimental procedure of a gasification experiment with steam and carbon dioxide is rather similar to the pyrolysis run described in the previous subsection. The main difference is the use of the steam furnace (cf. Figure 5.1) instead of the high temperature furnace and a slightly different temperature program. Further, experiments with the steam furnace have a bigger cycle time and are complicated by the operation of the steam generator and additional heaters that take longer to reach thermal equilibrium.

The temperature program used for the TG experiments with reactive H_2O-CO_2 sample environments consists of a dynamic ramp between 470 and 1250 K with a constant heating rate, $\beta = 20$ Kmin^{-1}. The faster heating rate compared to the pyrolysis runs was chosen for the sake of efficiency.

The following gas flows were used, depending on the desired reactive gas composition: The purge gas flow through the balance was maintained at 153 ml$_N$/min Ar during all experiments. Argon did not enter the furnace chamber but needed to be subtracted in the calculation of the product concentrations. The reactive gas flow consisted of the three components

- argon,
- steam,
- carbon dioxide

at a respective flow rate that varied between 0 and 100% of the total reactive gas flow rate. For material safety reasons, the reactive gas stream was kept below the 153 ml$_N$/min argon of the purge gas stream. The pressure inside the furnace chamber was atmospheric in all experiments.

The complete list of the H_2O-CO_2 gasification experiments with the respective temperature and concentration data is presented in Table 5.3.

The steam flow was controlled by a temperature safety valve and started only when the furnace reached a minimum temperature of 150 °C, which is achieved during the automatic preheating phase of the furnace. At the start temperature of the experiment, the reactive gases are still inert vis-à-vis coke. When the thermobalance finished the automatic preheating sequence from ambient temperature to the programmed start point at 470 K, both reactive gases, steam and carbon dioxide were present without reacting with the coke, and the experiment is started and completed automatically.

Since the off-gas contained a considerable amount of water depending on the chosen composition of the reactive gas, a cooler was used after the ther-

Table 5.2: Pyrolysis runs in the high temperature furnace with the corresponding experimental parameters.

coke type		m_0 (mg)	$T_0/\beta/T_{end}$ (K, K/min)	purge F_{Ar} (ml_N/min)	reactive gas $F_{r.g.}$	reactive gas H_2O-CO_2 (%-%)
Flexicoke	200-250	41.7	330/10/1823	153	0	0-0
PD coke	200-250	34.6	330/10/1823	153	0	0-0

Table 5.3: Reactive gasification runs in the steam furnace with the corresponding experimental parameters.

coke type	m_0 PD (mg)	m_0 Flexi (mg)	$T_0/\beta/T_{end}$ (K, K/min)	purge F_{Ar} (ml_N/min)	reactive gas $F_{r.g.}$	reactive gas H_2O-CO_2 (%-%)
	40.4	40.3	473/20/1523	153	145	0-20
	40.9	40.5	473/20/1523	153	145	0-40
	40.4	40.2	473/20/1523	153	145	0-60
	40.7	40.7	473/20/1523	153	145	0-80
	40.2	40.1	473/20/1523	153	145	0-100
	40.3	40.3	473/20/1523	153	145	20-0
Flexicoke	40.4	40.7	473/20/1523	153	145	40-0
250-355	40.4	40.0	473/20/1523	153	145	60-0
	40.5	40.8	473/20/1523	153	145	80-0
PD coke	40.2	40.3	473/20/1523	153	145	100-0
250-355	40.3	40.6	473/20/1523	153	145	80-20
	40.2	40.6	473/20/1523	153	145	60-40
	40.7	40.5	473/20/1523	153	145	40-60
	40.7	40.2	473/20/1523	153	145	20-80
	40.6	40.8	473/20/1523	153	145	20-20
	40.7	40.3	473/20/1523	153	145	40-20
	40.6	40.6	473/20/1523	153	145	60-20

mobalance to condensate excess steam. This had to be done to protect the micro GC from contamination with liquid water. The water concentration in the off-gas could therefore not be measured directly and was calculated by means of an oxygen mass balance.

5.2 Results

Based on the results of the proximate analysis, the coke is subdivided in volatiles, fixed carbon, and ash (cf. Table 2.2 on page 19 showing the respective amounts of Flexicoke and PD coke). Both the volatiles and the fixed carbon fractions are assumed to participate in the gasification reaction, but only the volatiles are subjected to pyrolysis. The total weight loss of the coke sample is defined as the sum of the pyrolysis and gasification weight loss,

$$X = X_G\left(1 - X_P\right) + X_P \qquad (5.1)$$

where

$$X(T) = 1 - \frac{m(T)}{m_0} \qquad (5.2)$$

X_P and X_G are the conversions due to pyrolysis and H_2O-CO_2 gasification, respectively ($X_G \in [0,1]$, $X_P \in [0, \sum_i x_i]$). Adjustment due to the ash content is not necessary because it represents only 0.33% weight for PD coke and 0.40% weight for Flexicoke. Thus, the measured rate of gasification is based on the weight loss data recorded by the TG:

$$-r_C = (1 - X_G)^{-2/3} \frac{dX_G}{dT}\beta \qquad (5.3)$$

Since pyrolysis and H_2O-CO_2 gasification occur simultaneously, a set of experiments using only inert gas were performed, and the resulting weight loss curve was subtracted from that obtained using reactive gases H_2O and CO_2 for calculating the rate of the H_2O-CO_2 gasification.

5.2.1 Pyrolysis

The TG and DTG curves obtained for the pyrolysis of Flexicoke and PD coke are shown in Figure 5.3. The experimental runs were performed with 200-250 μm particles, subjected to an argon flow of 210 ml/min and a linear heating rate of 10 Kmin^{-1} in the 400-1750 K temperature interval.

According to the kinetic model elaborated in Chapter 4.3, pyrolysis is considered as a linear combination of first order decomposition steps. The

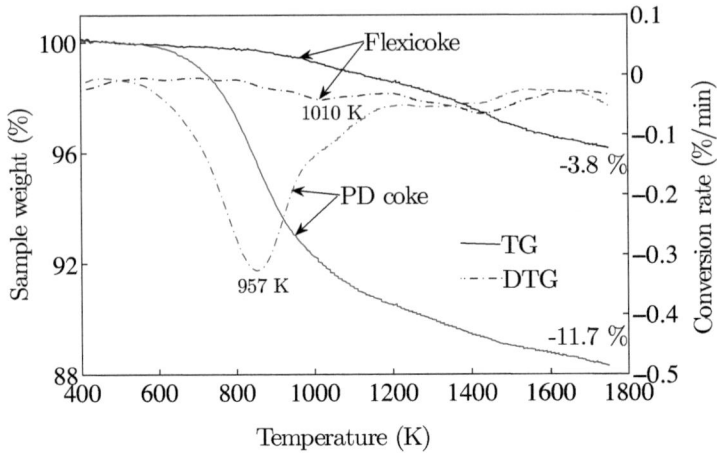

Figure 5.3: TG and DTG data for the pyrolysis of Flexicoke and PD coke. $d_P = 200\text{-}250$ μm, $\beta = 10$ Kmin^{-1}.

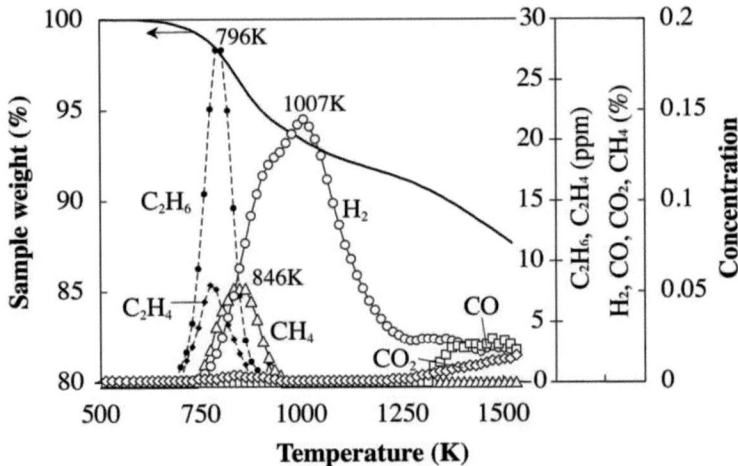

Figure 5.4: Weight loss and product gas composition as a function of temperature for the pyrolysis of PD coke. $d_P = 250\text{-}355$ μm, $\beta = 10$ Kmin^{-1}.

5.2. RESULTS

complete decomposition of a pseudo component is characterized by a minimum in the DTG curve. Both cokes indicate pyrolysis in a two-step process: 1) the DTG curves of Flexicoke and PD coke exhibit a first minimum at 1010 and 957 K, respectively. Thereby, the first weight loss step of PD coke is remarkably bigger than that of Flexicoke, indicating a higher amount of volatile components. 2) between 1300 and 1400 K, both Flexicoke and PD coke exhibit a second minimum. The total weight loss during pyrolysis up to 1750 K is 3.8% for Flexicoke and 11.7% for PD coke.

Figure 5.4 shows the off-gas analysis of a pyrolysis run with PD coke (d_P = 250-355 μm) in the steam furnace. This experiment is used in the following section to calculate the rate of the H_2O-CO_2 gasification of PD coke using Equation 5.1. Gases evolved during the first step of the pyrolysis of PD coke include a C_2H_4 peak at 773 K, C_2H_6 at 796 K, CH_4 at 846 K, CO_2 at 839 K, and H_2 at 1007 K. At temperatures above 1250 K the off-gas also contained small amounts of CO and CO_2. The results for PD coke are similar to those typically observed during pyrolysis of coal and other types of coke [41, 43].

5.2.2 H_2O-CO_2 gasification

TG and evolved gas analysis. Figures 5.5 and 5.6 show the TG curves and the evolved gas analysis as a function of the sample temperature for the steam gasification of Flexicoke and PD coke. The TG experiments were performed with 250-355 μm particles subjected to a 60% H_2O-Ar reactive atmosphere and a linear heating rate of 10 Kmin^{-1} in the 500-1520 K temperature interval.

The TG curve for the reactive gasification of Flexicoke (Figure 5.5) is not affected by the pyrolysis reaction, which is essentially negligible. In the case of PD coke (Figure 5.6), however, the TG curve is the result of a superposition of pyrolysis and reactive gasification. Pyrolysis of PD coke takes place at temperatures between 650 and 1000 K, resulting in a weight loss of about 10% and coming along with the characteristic pyrolysis gases in the evolved gas (cf. Chapter 5.2.1). Common to both Flexicoke and PD coke, the reactive gasification with steam proceeds at higher temperatures, starting at about 1000 K and reaching essentially full conversion at sufficiently high temperatures and reaction times.

Besides the pyrolysis gas products already considered in Figure 5.4, the main gasification products of petcoke are H_2, CO, and CO_2. At above 1210 K, steam reacts with the sulfur present in the petcoke to H_2S. CH_4 concentration

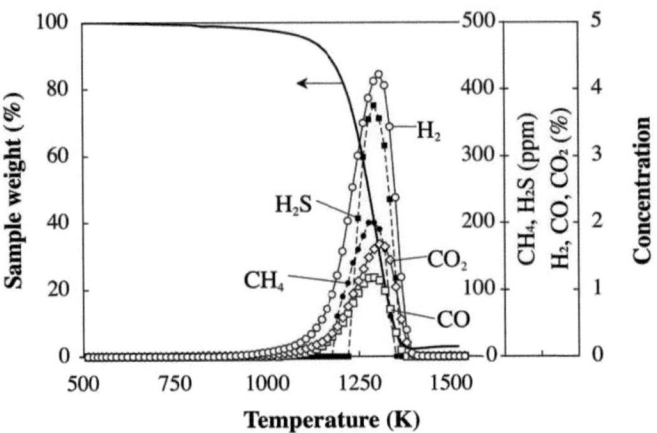

Figure 5.5: Weight loss and product gas composition as a function of temperature for the steam gasification of Flexicoke.
$p_{H_2O} = 0.6$ bar, $d_P = 250\text{-}355$ μm, $\beta = 10$ Kmin^{-1}.

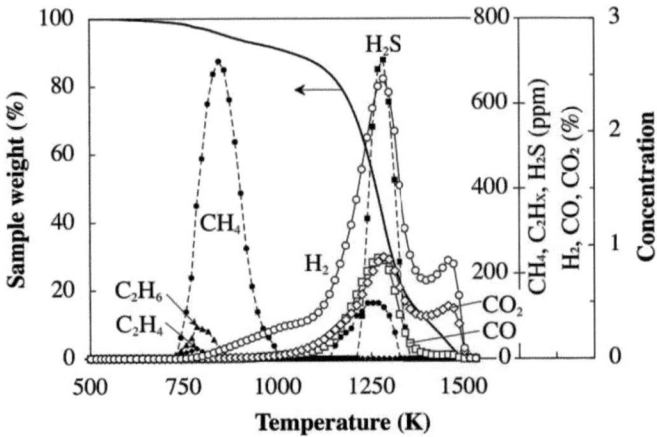

Figure 5.6: Weight loss and product gas composition as a function of temperature for the steam gasification of PD coke.
$p_{H_2O} = 0.6$ bar, $d_P = 250\text{-}355$ μm, $\beta = 10$ Kmin^{-1}.

5.2. RESULTS

is less than 0.013% for PD coke and less than 0.02% for Flexicoke.

The steam gasification of PD coke further appears to proceed following two regimes, a faster one up to 1310 K, and a slower one at higher temperatures. The significant slow down of the gasification rate at $\approx 1300 K$ is explained with a thermally induced deactivation process causing a loss of reactive surface (cf. Figures 6.11, 6.12, and 6.13 in Chapter 6.1 showing the effect of temperature on the composition and the reactive surface of PD coke).

Effect of the reactive gas concentration. The effect of the steam and CO_2 concentration on the rate of the reactive gasification is shown in Figures 5.7 and 5.8 for Flexicoke and PD coke, respectively. Petcoke samples with 250-355 μm particles were gasified in binary H_2O-Ar and CO_2-Ar atmospheres with a reactive gas concentration in the 0-100% H_2O/CO_2 concentration range and a linear heating rate of 20 $Kmin^{-1}$ in the 500 to 1520 K temperature interval.

As already mentioned, the simultaneous occurrence of pyrolysis clearly affects the H_2O-CO_2 gasification of PD coke, whereas it is negligible in the case of Flexicoke. At the same time, the rate of the steam gasification reaction is significantly faster than the rate of the CO_2-gasification reaction for both coke types. The observed difference is bigger for PD coke that did not reach full conversion in dry gasification runs with CO_2 partial pressures up to 1 bar and temperatures up to 1520 K. At higher concentration of the reactive gas, the gasification exhibits a saturation behavior typical for heterogeneous gas-solid reactions. A thermally-induced deactivation at above 1310 K is observed for the steam gasification of PD coke.

Effect of the particle size. The effect of the particle size on the rate of the reactive gasification in a 60% H_2O-Ar atmosphere is shown in Figures 5.9 and 5.10 for Flexicoke and PD coke, respectively. The samples were heated with a linear heating rate of 10 $Kmin^{-1}$ from 500 to 1520 K.

In the case of Flexicoke, the effect of the particle size on the rate of the reactive gasification was negligible. However, the gasification rate of PD coke showed a significant dependence on the particle size, and the reaction rate increased with decreasing mean particle size. The fastest rates were observed for the powder size samples with a mean particle size ($d_{P,50}$) of 1.2, 3.8, and 26 ('> 80') μm (cf. Table 2.3 on page 21 with the respective sample properties).

Pyrolysis of PD coke starts at somewhat lower temperatures (about 600 K) and has a higher reaction extent for powder size fractions obtained from

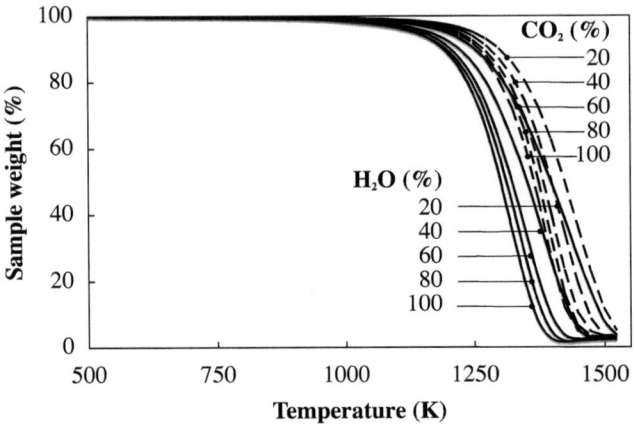

Figure 5.7: Weight loss as a function of temperature for the gasification of Flexicoke in binary H_2O-Ar and CO_2-Ar mixtures. $d_P = 250\text{-}355$ μm, $\beta = 20$ Kmin^{-1}.

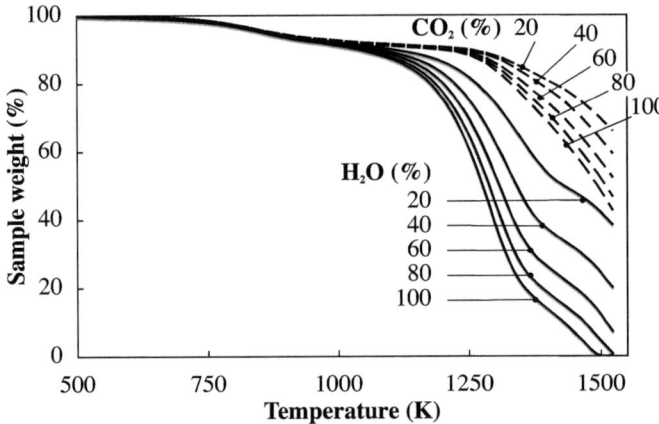

Figure 5.8: Weight loss as a function of temperature for the gasification of PD coke in binary H_2O-Ar and CO_2-Ar mixtures. $d_P = 250\text{-}355$ μm, $\beta = 20$ Kmin^{-1}.

5.2. RESULTS

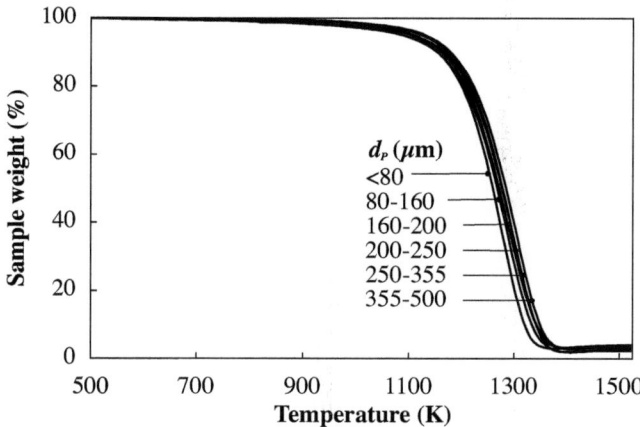

Figure 5.9: Weight loss as a function of temperature for the steam gasification of Flexicoke. The parameter is the mean particle size. $p_{H_2O} = 0.6$ bar, $\beta = 10$ Kmin^{-1}.

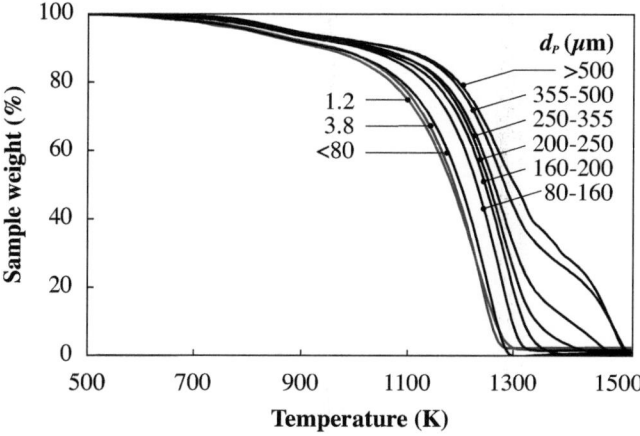

Figure 5.10: Weight loss as a function of temperature for the steam gasification of PD coke. The parameter is the mean particle size. $p_{H_2O} = 0.6$ bar, $\beta = 10$ Kmin^{-1}.

grinding with the jet mill and the ball mill. For all particles bigger than 200-250 µm, the reaction does not reach completion before below 1310 K, and its rate undergoes attenuation by thermal deactivation.

The averaged weight of the residuum found at the end of the TG experiment was 3.6 and 1.1 wt% for Flexicoke and PD coke, respectively.

Experimental particle effectiveness

The particle effectiveness factor, η_P, is determined experimentally using Equation (4.15) from the particle model presented in Chapter 4.1.1:

$$\eta_P = \frac{r_{obs}}{r_{intr}}$$

η_P is the ratio of the observed overall rate of reaction, r_{obs}, and the intrinsic rate of reaction, r_{intr}, that would be observed under the same conditions if diffusion in the solid was instantaneous.

The intrinsic rate, r_{intr}, is obtained experimentally from gasification experiments using very small particles since the diffusion resistance inside the particle becomes negligible for $R_P \to 0$. Alternatively, r_{intr} is accessible via the extrapolation of a series of experiments using particles of different sizes to $R_P = 0$, which is done in the following.

Figures 5.11 and 5.12 show experimentally measured rate data for five temperatures between 1050 and 1250 K. The radii of the coke samples used for the experiments are in the 20-213 µm and 0.6-372 µm range for Flexicoke (Figure 5.11) and PD coke (Figure 5.12), respectively. The dashed lines represent the extrapolation of the experimental rate data at each temperature level to $R_P = 0$ in order to obtain the intrinsic rate, r_{intr}. Once the intrinsic rate at a certain temperature is known, the effectiveness factor, η_P, is readily calculated by application of Equation (4.15) to any rate measured at the same temperature with particles having a different size.

Figure 5.13 shows the effectiveness factors calculated for the experiments in Figures 5.9 and 5.10. As expected, η_P of samples consisting of small particles is one. $\eta_P = 1$ is observed for Flexicoke particles with $R_P < 10^{-5}$ m and for PD coke particles with $R_P < 10^{-6}$ m, whereas PD coke with $R_P = 372$ µm has an effectiveness as low as 20%. Figures 5.11, 5.12 and 5.13 further show that the effectivity of PD coke is stronger affected by an increase of the particle size than Flexi coke. An effect of the particle temperature, which is known to have a significant influence on the particle effectiveness due to the exponential temperature dependence of the reaction rate, was not found in the investigated temperature interval. An extension

5.2. RESULTS

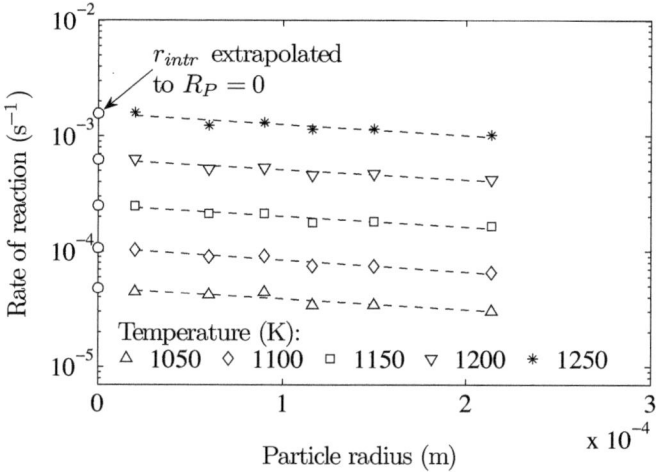

Figure 5.11: Calculation of r_{intr} ('○') for Flexicoke via extrapolation of isothermal rate data to $R_P = 0$.

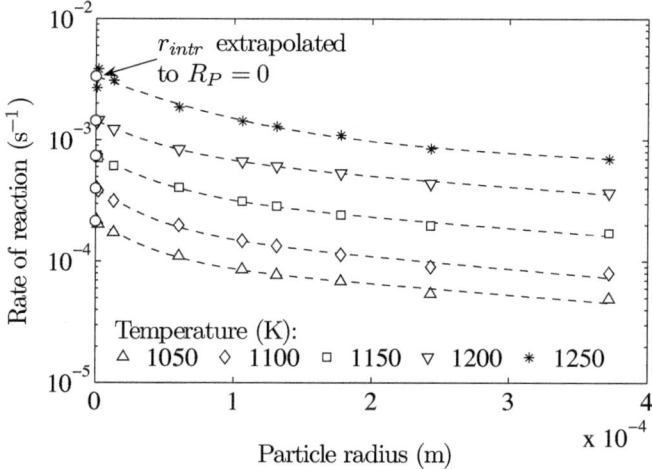

Figure 5.12: Calculation of r_{intr} ('○') for PD coke via extrapolation of isothermal rate data to $R_P = 0$.

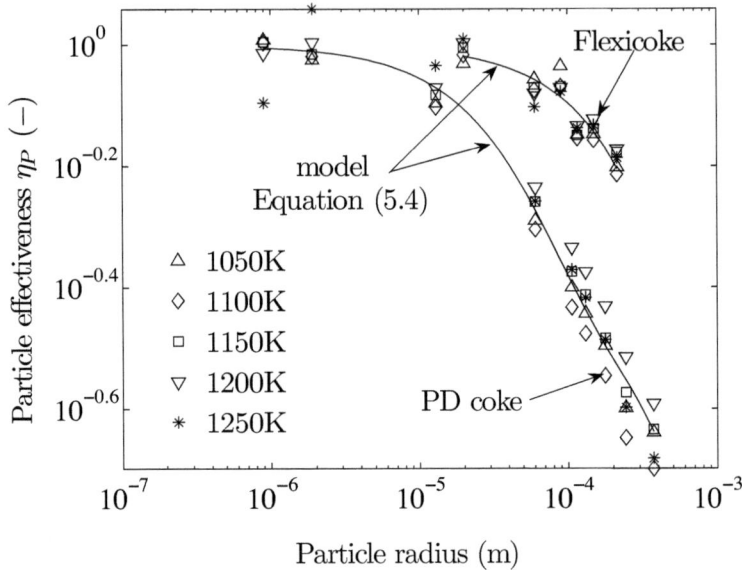

Figure 5.13: Experimental particle effectiveness factor η_P for the steam gasification of Flexicoke and PD coke.

of the temperature interval would certainly yield a different result, but this was not feasible with the setup at hand.

To calculate the rate of the reactive gasification in the chapters that follow, the experimental effectiveness shown in Figure 5.13 is correlated with a sum of exponentials according to

$$\eta_P = a\exp(-bx) + c\exp(-dx) \tag{5.4}$$

The parameters in Equation (5.4) are calculated for the additional constraint $a + c = 1$ to force the curve through the point $(0,1)$ since $\eta_P = 1$ for $R_P = 0$. Numerical values for the parameters a, b, c and d are presented in Table 5.4, and the calculated fit is shown in Figure 5.13 as a solid line together with the experimental data.

5.2. RESULTS

Table 5.4: Numerical values for the parameters a, b, c and d in Equation (5.4)

	a	b	c	d
Flexicoke	$2.188 \cdot 10^{-3}$	$2.048 \cdot 10^4$	$9.978 \cdot 10^{-1}$	$2.281 \cdot 10^3$
PD coke	$5.709 \cdot 10^{-1}$	$2.584 \cdot 10^4$	$4.291 \cdot 10^{-1}$	$1.913 \cdot 10^3$

Thermal deactivation

The evaluation of the gasification experiments with PD coke showed that only rate data collected at temperatures below $\approx$ 1300 K could be used for the calculation of the H_2O-CO_2 gasification kinetics. At temperatures above 1300 K, a significant reduction of the reaction rate was observed, even while further increasing the temperature.

Figure 5.14 shows the gasification rates of PD coke for experiments with binary CO_2-Ar reactive gas (cf. Chapter 5.2.2) and ternary H_2O-CO_2-Ar reactive gas (cf. Chapter 5.3.2). As previously discussed, the rate of the H_2O-CO_2 reactive gasification is superimposed by the rate of the pyrolysis reaction causing a peak at 850 K. Gasification starts at about 900 and 1200 K for H_2O and CO_2, respectively, and exhibits an exponential growth towards 1300 K, where a maximum occurs.

The location of maximum is not the same for all experiments. The reactive gas composition has an effect on the location of the rate peak observed before deactivation takes place. In experiments with high reactive gas concentration (especially steam) and high reaction rates, the temperature of the maximum is found to be close to 1300 K. In experiments with CO_2-Ar mixtures having a considerably lower reactivity with PD coke than steam the maximum is observed at temperatures around 1380 K.

The corresponding temperature of the peak rate in Figure 5.14 is well correlated with an exponential function according to

$$r_{peak} = a \exp(b \cdot T_{peak}) \tag{5.5}$$

with $a = 1.153 \cdot 10^9$ and $b = -0.01934$. The R squared of the fit is $R^2 = 0.981$.[2] The peak values of the rate are shown together with the exponential fit in the enlarged clipping of Figure 5.14. The deactivation process can be considered as a chemical reaction by itself, which is competing with

---
[2]The coefficient of determination, R^2, is defined as the ratio of the sum of squares

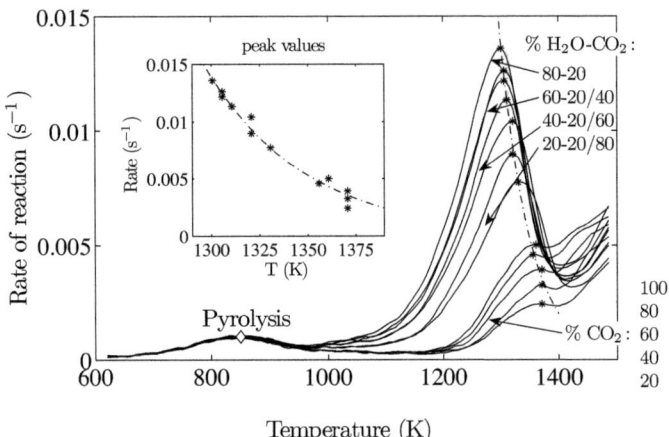

Figure 5.14: Thermal deactivation of PD coke above 1300 K during gasification experiments with binary CO_2-Ar and ternary H_2O-CO_2-Ar reactive gas. $X_C < 1$, $d_P = 250\text{-}355$ μm, $\beta = 20$ Kmin^{-1}

the reactive gasification. The temperature corresponding to the maximum deactivation rate for a specific composition of the reactive gas is considered as the characteristic deactivation temperature. This deactivation temperature is found at the second inflection point of the gasification rate, about 20 K above the peak temperature of the reactive gasification.

The thermal deactivation of PD coke is further investigated by means of an additional set of gasification runs in the TG using PD coke samples pretreated at different final pyrolysis temperatures, T_P. The respective TG and DTG date is shown in Figures 5.15 and 5.16. Pretreatment is done using PD coke samples with 250-355 μm particle diameter. 40 mg samples are pyrolyzed in argon using a 20 Kmin^{-1} heating rate and different final pyrolysis temperatures $T_P \in [973, 1773]$ K. After pyrolysis, the samples are gasified in 60% H_2O-Ar and using a $\beta = 10$ Kmin^{-1} heating rate.

Figure 5.15 shows the TG curves of untreated (dashed line) samples and explained by the regression, SS_R, to the total sum of squares, SS_Y [80].

$$R^2 = \frac{SS_R}{SS_Y} \tag{5.6}$$

5.2. RESULTS

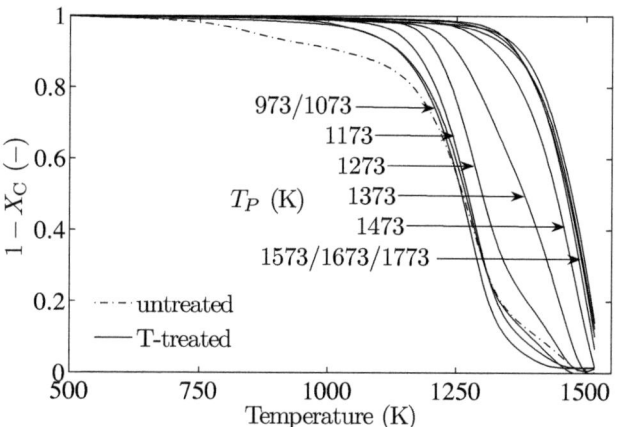

Figure 5.15: Weight loss as a function of temperature for the gasification of thermally pretreated PD coke.
$d_P = 250\text{-}355$ μm, $p_{H_2O} = 0.6$ bar, $\beta = 10$ Kmin^{-1}

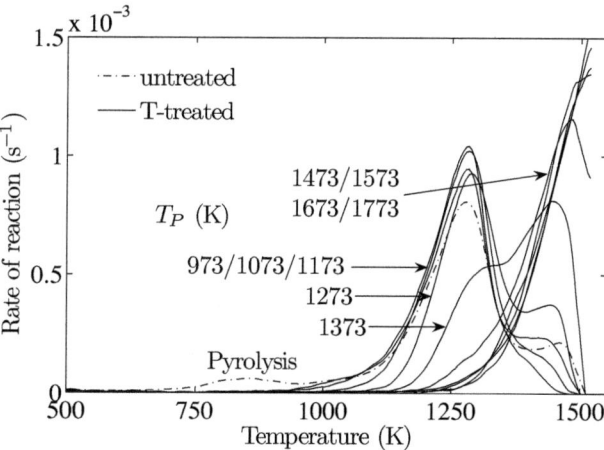

Figure 5.16: Rate of reaction as a function of temperature for the gasification of thermally pretreated PD coke.
$d_P = 250\text{-}355$ μm, $p_{H_2O} = 0.6$ bar, $\beta = 10$ Kmin^{-1}

samples that were thermally treated (solid lines) at a maximum pyrolysis temperature, T_P. Two effects can be observed:

1. Temperature pretreatment leads to an anticipation of the pyrolysis step in the 700-900 K range.

2. The TG curves of samples pretreated at pyrolysis temperatures in the 1273-1473 K interval show a shift of the TG curve up to 200 K towards higher temperatures.

The rates of the gasification reaction presented in Figure 5.16 show a maximum at 1280 K as long as the pretreatment temperature did not exceed the deactivation threshold temperature at about 1310 K. Maintaining the reactive gas composition and the heating rate constant, the corresponding temperature of the peak rate is the same for all runs, independent of carbon conversion.

In experiments with Flexicoke, no thermal deactivation was observed at any temperature. It is assumed that the more severe conditions in the production process of Flexicoke result in a product being less affected by subsequent temperature treatment. Flash pyrolysis experiments in H_2O-Ar in the solar furnace at PSI showed a possibility to avoid thermal deactivation also for PD coke.

TG and DTG error estimation

The detection limit for weight changes is affected by the background noise of the TG signal and the accuracy of the electronic balance. To measure a significant rate of reaction, the underlying weight change has to be bigger than the background noise of the TG signal. As a measure for the background noise, the standard deviation

$$\sigma_x = \sqrt{\frac{1}{N} \sum_{i=1}^{N} (x_i - \bar{x})^2} \qquad (5.7)$$

of the TG signal before and after applying a low pass filter[3] is used. Figure 5.17 shows the noise for the experimental data presented in Chapters 5.2.1 and 5.2.2. The calculated amplitudes $|x - \bar{x}|$ are neither depending on temperature nor on conversion, which is a strong function of temperature (in the high temperature range, conversions of 100% have been obtained).

---

[3]The applied filter is a locally weighted scatter plot smooth using least squares quadratic polynomial fitting

5.3. KINETIC ANALYSES

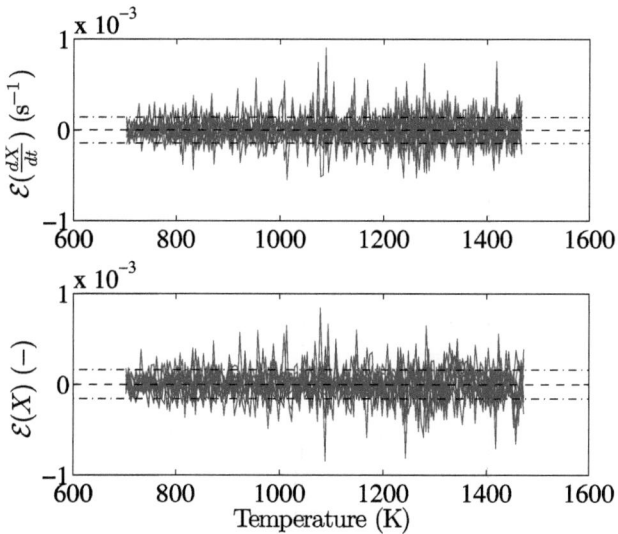

Figure 5.17: Experimental error (noise) of the TG and DTG curves during the pyrolysis and gasification runs. The dashed line represents the arithmetic mean, $\bar{x}$, and the dash-dotted line is $\pm\sigma_x$.

The standard deviation of the raw signal with respect to the smoothed signal was calculated for the TG ($\sigma(X) = 1.59 \cdot 10^{-4}(-)$) and the DTG curves ($\sigma\left(dX/dt\right) = 1.44 \cdot 10^{-4}$ (s^{-1})). Those values are considered as the lower detection limit of the thermobalance.

5.3 Kinetic analyses

5.3.1 Pyrolysis

Figures 5.18 and 5.19 show the measured (data points) and modeled (curves) conversion rate dX_P/dt obtained for the pyrolysis of Flexicoke and PD coke, respectively. A smoothing algorithm based on least squares quadratic polynomial fitting is applied to the measured TG curve. The kinetic parameters k_0, E_A, and the maximum conversion x_i for two pseudo-components, defined by Equations (4.34), (4.35) and (4.36), were determined by the least-square

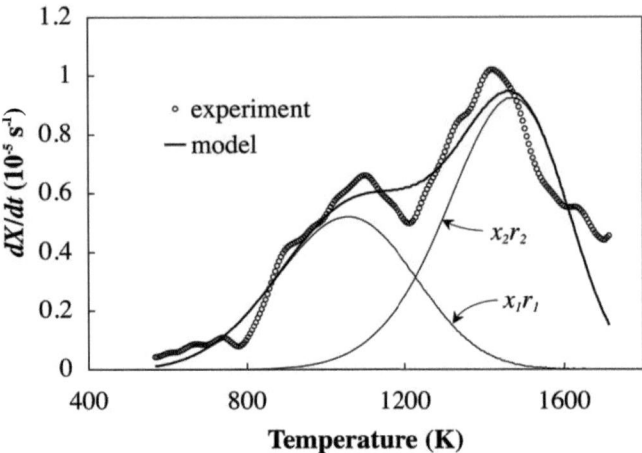

Figure 5.18: Measured (data points) and modeled (curves) conversion rate dX_P/dt obtained for the pyrolysis of Flexicoke.

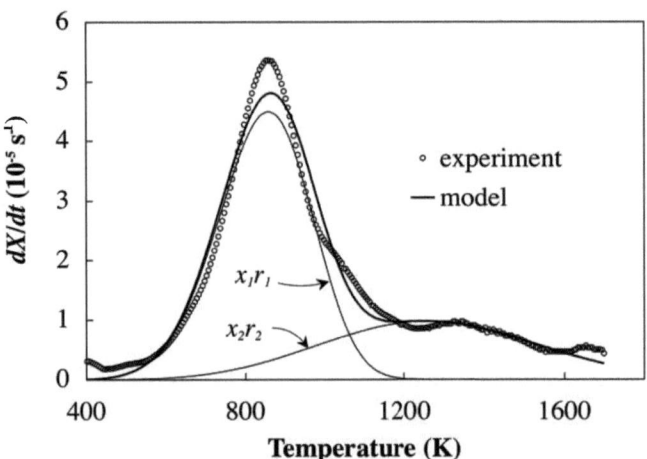

Figure 5.19: Measured (data points) and modeled (curves) conversion rate dX_P/dt obtained for the pyrolysis of PD coke.

5.3. KINETIC ANALYSES

Table 5.5: Kinetic parameters for the pyrolysis of petroleum coke.

		k_0 (s^{-1})	E_A (J/mol)	x_i (−)
Flexicoke	r_1	4.152·10^{-2}	5.157·10^4	0.0136
	r_2	1.881·10^0	1.210·10^5	0.0207
PD coke	r_1	1.905·10^{-1}	5.087·10^4	0.0794
	r_2	7.528·10^{-3}	4.626·10^4	0.0393

method. The root mean square (RMS) of the absolute error between the experimentally measured and the theoretically modeled reaction rate dX_P/dt was minimized with an algorithm of MATLAB [37]. The kinetic parameters are listed in Table 5.5 for Flexicoke and PD coke.

The weight losses of Flexicoke for the first and the second reactions are 1.36 and 2.07%, respectively, giving a total of 3.43%. The weight losses of PD coke for the first and the second reactions are 7.94 and 3.93%, respectively, giving a total mass loss of 11.87%. The RMS of the absolute error between the measured and the modeled petcoke pyrolysis is $8.40 \cdot 10^{-7}$ for Flexicoke and $2.24 \cdot 10^{-6}$ for PD coke.

5.3.2 H$_2$O-CO$_2$ gasification

The calculation of the Arrhenius parameters for the reactive gasification in H$_2$O-CO$_2$ mixtures is based on the intrinsic rate, r_{intr}, not affected by mass transfer limitations inside the particle. r_{intr} is calculated with Equation (4.15)

$$r_{intr} = \frac{r_{obs}}{\eta_P}$$

where η_P is the particle effectiveness factor calculated with Equation (5.4) and the data from Table 5.4, presented in the previous section. The overall rate of the reactive gasification, r_{obs}, is obtained by subtraction of the pyrolysis weight loss, X_P, from the overall weight loss, X, according to Equation (5.1) and taking the time derivative.

The Arrhenius parameters of the rate constants defined by Equation (4.53) (derived by applying the *oxygen-transfer* mechanism) and Equation (4.83) (derived by applying the *extended* mechanism) were determined numerically using the same algorithm by MATLAB [37] for minimizing the RMS error between measured and modeled data as for the pyrolysis kinetics. The numerical procedure is carried out simultaneously for all temperatures

Table 5.6: Arrhenius parameters of the rate constants derived by applying the oxygen-exchange mechanism [Equation (4.53)] for the H_2O-CO_2 gasification of PD coke. RMS = $3.80 \cdot 10^{-5}$.

	k'_1 ($Pa^{-1}s^{-1}$)	k'_2 ($Pa^{-1}s^{-1}$)	k'_3 (−)
k_0 (k'_i)	$4.856 \cdot 10^3$	$5.601 \cdot 10^0$	$1.712 \cdot 10^2$
E_A (J/mol)	$2.851 \cdot 10^5$	$1.866 \cdot 10^5$	$9.929 \cdot 10^4$

and all combinations of partial pressures (H_2O-CO_2-Ar and CO_2-Ar mixtures) applied in the experiments of Table 5.3. Table 5.6 lists the best-fit kinetics parameters of the rate constants derived for the oxygen-exchange mechanism of the steam-gasification of PD coke. The results are in agreement with data presented in gasification studies with pure H_2O or CO_2 [47]. Further, the gasification with H_2O-CO_2 mixtures can be described using the rate constants obtained from experiments with pure gases, as claimed previously [21].

As far as the steam gasification of Flexicoke is concerned, the oxygen-exchange mechanism is not capable of describing with reasonable accuracy the reaction rate. As it will be shown in the following Figures, the addition of CO_2 to a reacting gas containing more than 60 % H_2O does not result in an appreciable increase of the reaction rate, as predicted by Equation (4.53). Presumably, mass transfer is limited by slow diffusion of the reactive gas into the interior of the particles through their pores [21].

Table 5.7 lists the best-fit kinetic parameters derived for the extended mechanism of the H_2O-CO_2 gasification of Flexicoke and PD coke. Figure 5.20 shows the comparison of the experimentally measured (data points) and the theoretically modeled (curves) rate of reaction, $-r_C$, as a function of temperature for various H_2O-CO_2 mixtures. The parameter is the composition of the gaseous reactant, expressed in %H_2O-%CO_2 at 1 bar total pressure.

The extended model is closely obeyed over the 900-1300 K temperature range and over the complete range of H_2O/CO_2 concentrations. The root mean square (RMS) of the absolute error between the experimental and the modeled rate of reaction is $2.31 \cdot 10^{-5}$ s^{-1} for Flexicoke and $3.80 \cdot 10^{-5}$ s^{-1} for PD coke. Taking into account the different empirical particle effectiveness being 0.709 and 0.334 for 250-355 μm Flexicoke and PD coke, respectively, the overall rate of the steam gasification is essentially the same for both coke types, whereas the overall rate of the CO_2 gasification is considerably faster

Table 5.7: Arrhenius parameters of the rate constants derived by applying the extended mechanism [Equation (4.83)] for the H_2O-CO_2 gasification of Flexicoke and PD coke.

		K_1 ($Pa^{-1}s^{-1}$)	K_2 ($Pa^{-1}s^{-1}$)	K_4 (Pa^{-1})	K_5 (Pa^{-1})
Flexicoke[a]	k_0 (K_i)	$9.650 \cdot 10^0$	$3.364 \cdot 10^1$	$1.576 \cdot 10^{-5}$	$5.560 \cdot 10^{-4}$
	E_A (J/mol)	$2.132 \cdot 10^5$	$2.170 \cdot 10^5$	$8.276 \cdot 10^3$	$5.739 \cdot 10^5$
PD coke[b]	k_0 (K_i)	$1.158 \cdot 10^3$	$4.978 \cdot 10^{-1}$	$4.033 \cdot 10^{-8}$	$8.152 \cdot 10^{-6}$
	E_A (J/mol)	$2.707 \cdot 10^5$	$1.615 \cdot 10^5$	$7.068 \cdot 10^3$	$4.551 \cdot 10^2$

[a] RMS = $2.305 \cdot 10^{-5}$
[b] RMS = $3.796 \cdot 10^{-5}$

for Flexicoke than for PD coke.

Figure 5.21 is a cross-plot of the previous figure at 1273 K and shows a contour of reaction rates, $-r_C$ (s^{-1}), as a function of the H_2O and CO_2 partial pressures, for Flexicoke (a) and PD coke (b). The figure also shows the different reactive gas compositions used for the TG experiments ('⊗') and can be used to predict the effect of a change in the reactive gas composition on the intrinsic reaction rate. For example, adding CO_2 to a H_2O-CO_2 reactive gas at $p_{H_2O} > 0.7$ bar does not result in an increase of the reaction rate for Flexicoke, whereas it accelerates the reaction rate of PD coke assuming the same conditions.

At 1273 K, the intrinsic reaction rate in 100% H_2O at 1 bar is $6.5 \cdot 10^{-3}$ s^{-1} for PD coke, almost twice of the value observed for Flexicoke. Taking into account the particle effectiveness of a 300 µm sample (i.e. 0.709 for Flexicoke and 0.334 for PD coke) results in an overall reaction rate that is about the same for both coke types. Despite the differences in the values of the kinetic parameters K_1 and K_4 associated with the CO_2 terms of Equation (4.83), the intrinsic reaction rate with 100% CO_2 is similar for both types of coke, about $1.0 \cdot 10^{-3}$ s^{-1}.

5.4 Summary and conclusions

Kinetic rate laws for the combined pyrolysis-gasification of petcoke and determined experimentally the Arrhenius parameters were derived by dynamic

(a)

(b)

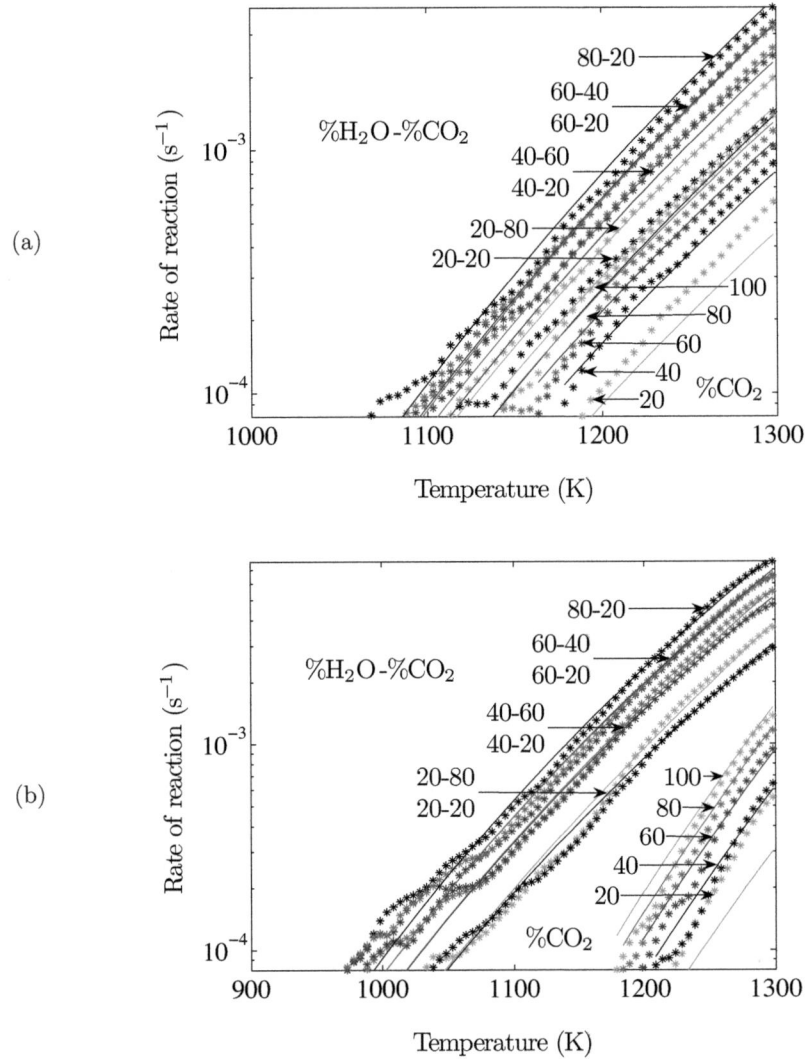

Figure 5.20: Experimentally measured (data points) and modeled (curves) reaction rate, $-r_\mathrm{C}$ (s^{-1}), as a function of temperature for the gasification of (a) Flexicoke and (b) PD coke. The parameter is the composition of the reactive gas expressed in %H$_2$O-%CO$_2$. The overall pressure is 1 bar.

5.4. SUMMARY AND CONCLUSIONS

(a)

(b)

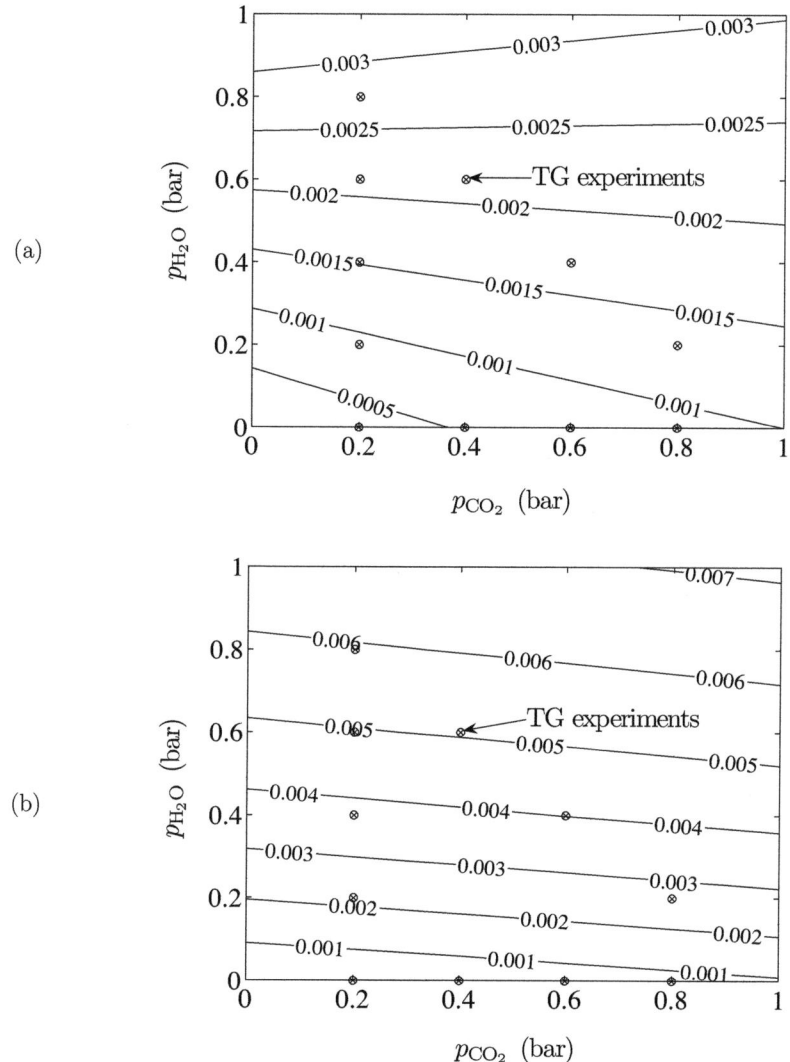

Figure 5.21: Reaction rate, $-r_C$ (s^{-1}), as a function of the H$_2$O and CO$_2$ partial pressures at 1273 K for (a) Flexicoke and (b) PD coke. The total pressure is 1 bar. '⊗' represents the conditions of the TG experiments.

thermogravimetric analyses of the Ar-pyrolysis and H_2O-CO_2-gasification of two types of Venezuelan petcokes (Flexicoke and PD coke). The pyrolysis is well described by applying a 2-pseudo-component decomposition mechanism. The gasification kinetics is modeled by applying two mechanisms: an oxygen-transfer mechanism based on reversible O-transfer surface reactions and irreversible carbon gasification, and an extended mechanism that considers, in addition, OH/H-sorption and H-transfer. The former, simpler, mechanism fails to describe the Flexicoke gasification, presumably because of mass transfer limitations inside the particle. The latter mechanism is able to simulate, with good experimental agreement, the H_2O-CO_2 gasification of both types of petcoke.

Steam gasification runs at considerably higher rates than dry gasification: The intrinsic reaction rate with 100% H_2O is $6.5 \cdot 10^{-3}$ s^{-1} for PD coke, almost twice the value observed for Flexicoke. The intrinsic reaction rate with 100% CO_2 is similar for both types of coke, about $1.0 \cdot 10^{-3}$ s^{-1}. Taking into account the differences in the empirical particle effectiveness being 0.709 and 0.334 for 250-355 μm Flexicoke and PD coke, the overall rate of H_2O gasification is comparable for both types of cokes, whereas the overall rate of CO_2 gasification is considerably faster for Flexicoke than for PD coke.

Main evolved gases during pyrolysis include C_2H_6, C_2H_4, CH_4, CO_2, CO, and H_2 for PD coke, and CO, CO_2, and H_2 for Flexicoke. Besides the pyrolytic gas products, the main gasification products are H_2, CO, CO_2, and H_2S for both cokes.

The dependence of the rate of the H_2O-CO_2-gasification on the particle size was modeled empirically by means of a particle size dependent effectiveness parameter, η_P, which is the ratio of the observed overall rate and the intrinsic rate in absence of mass transfer limitations. η_P is determined experimentally for both coke types.

Chapter 6

Experimental determination of the rate constants using lab scale integral reactors[1]

This chapter presents two series of steam gasification experiments performed in laboratory scale fluidized bed reactors. Two different reactor setups were used that differ with respect to the heat transfer mode from an external radiation source to the petcoke particles in the fluidized bed:

- The *direct irradiated fluidized bed* (DFB) reactor consists of a transparent quartz fluidization tube including a porous disk acting as a bed support. The reactants being coke and steam are heated by direct irradiation of the petcoke particles in the fluidized bed with concentrated thermal radiation from ETH's solar simulator.

- The *indirect irradiated fluidized bed* (IFB) reactor features an opaque fluidization tube, offering the possibility to use fluidization tubes made of different materials featuring different wall strength and optical properties. The heat transfer from the solar simulator to the reactants is indirect and occurs via heating of the fluidization tube and subsequent radiative and convective heat transfer from the wall to the bed.

---

[1]The experimental data presented in this chapter were obtained in the framework of two master theses at the Professorship in Renewable Energy Carriers at ETH Zurich, performed by M. Fasciana, F. Noembrini [22] (DFB experiments in Chapter 6.1) and R. Alvarez [2] (IFB experiments in Chapter 6.2). Material from this chapter has been published in 'D. Trommer, F. Noembrini, M. Fasciana, et al. Hydrogen production by steam gasification of petroleum coke using concentrated solar power - I. Thermodynamic and kinetic analyses. *International Journal of Hydrogen Energy*, 2005. 30(6): p. 605-618'.

The gasification experiments with a direct and an indirect irradiated fluidized bed reactor offer the possibility of analyzing if the heat transfer mode has an effect on the reaction kinetics. Because other experimental parameters including the properties of the radiation source, the fluidized bed parameters, and the type of fuel are the same for both reactors, the experimental configurations described above permit a direct investigation of a possible effect of the heat transfer mode on the reaction kinetics.

Chapter 6.1 presents the DFB setup and the experimental results obtained for the gasification using direct irradiation of the reactants. Chapter 6.2 discusses the IFB setup and the experimental results obtained for gasification using indirect irradiation. The gasification experiments with both setups are performed isothermally using different bed temperatures in the range of 1070-1405 K. The corresponding rate constants are calculated in Chapter 6.3 by solving the mass balance for every species over a differential reactor element and performing an Arrhenius analysis to derive the pre-exponential factors and activation energies.

6.1 Fluidized bed reactor with direct irradiation

6.1.1 Reactor setup

The chemical reactor used for the kinetic study using direct irradiation consisted of a 25 mm-outer diameter, 1.5 mm-thickness, 25 cm-height quartz tube containing a petcoke/steam fluidized bed directly exposed to concentrated thermal irradiation. Figure 6.2 shows a photograph of the quartz fluidized bed reactor (QFBR) before (left) and after (right) an experimental run.

The reactor concept was previously used to analyze the steam gasification of activated charcoal [63]. For the gasification experiments with coke, the reactor was added a new side access below the porous quartz disk acting as bed support. With this arrangement, the thermocouple enters the fluidized bed from the bottom via the side access and through a pin-hole in the frit. In opposition to the previous design featuring a top access, the thermocouple is now protected from the intense thermal radiation, what leads to an increased lifetime and a reduced failure rate.

Tests were conducted at the ETH's high-flux solar simulator - a high-pressure argon arc close-coupled to precision elliptical-trough mirrors - de-

6.1. DFB REACTOR

scribed in [36]. This research facility provides a rapid external source of intense thermal radiation that approaches the heat transfer characteristics of highly concentrating solar systems. Power, power fluxes (power per unit area), and temperatures can be adjusted to meet the specific requirements by simply varying the electrical input power to the arc electrodes.

Figure 6.3 shows a photograph of the experimental setup during an experiment. The quartz fluidized bed reactor is mounted on a mobile unit standing below the reflector of the solar simulator. The upstream steam supply and downstream quenching and filter units are located below a massive, water-cooled copper plate that protects the setup from the solar simulator's intense thermal radiation.

The tubular reactor is positioned at the focal plane of the solar simulator and subjected to a power flux irradiation of 18 W/cm^2 laterally and 130 W/cm^2 at the top. The nominal fluidized bed temperature is measured by means of a solar-blind pyrometer and by thermocouples type S submerged in the fluidized bed. Gas flows are controlled using Bronkhorst HI-TEC flow controllers. After exiting the reactor, the product gases are quenched to room temperature and analyzed on-line by gas chromatography (the GC configuration is identical with the one used for the TG experiments presented in Chapter 5). The flowchart in Figure 6.1 shows the experimental setup with its up and downstream components for steam supply and off-gas analysis, and Table 6.1 lists the bulk properties of the employed samples.

Table 6.1: Mean properties of petcoke particles used as reactants in this study.

	Flexicoke	PD coke
Particle size	250-355 μm	250-355 μm
Bulk density	0.74 g/cm^3	0.57 g/cm^3
BET specific surface	11 m^2	1 m^2
Cumulative pore volume	0.027 cm^3/g	0.0018 cm^3/g

6.1.2 Experimental procedure

During a typical experiment, a batch of 4 g solid reactants was first heated to the desired temperature under an argon flow and then subjected to continuous H_2O-Ar flow under isothermal conditions. Using gas flow rates in the range 0.8-2.0 l_N/min, a stable fluidized bed was obtained by placing 1-3 mm

86 CHAPTER 6. RATE CONSTANTS FROM INTEGRAL REACTORS

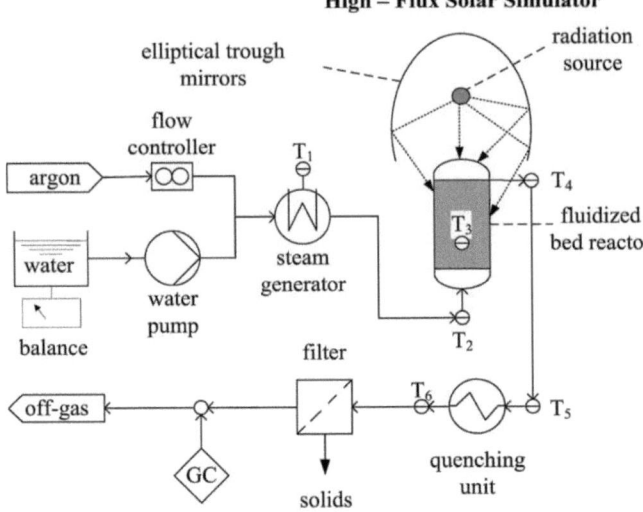

Figure 6.1: Flowchart of the fluidized bed reactor setup in the ETH's high-flux solar simulator.

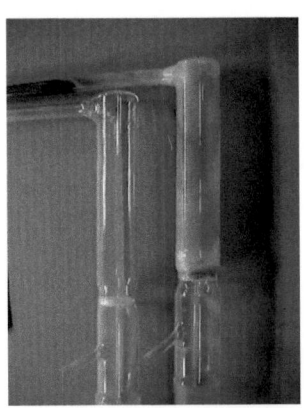

Figure 6.2: Quartz fluidized bed reactor before (left) and after (right) experiment.

Figure 6.3: The ETH's solar simulator with the quartz reactor during an experiment.

6.1. DFB REACTOR

Al_2O_3 granules at the bottom of the bed. The fluidized bed characteristics were: volume ≈ 12 cm^3, void fraction $\epsilon_v \approx 0.9$, superficial velocity $v_0 \approx$ 20-30 cm/s, $Re_P \approx 10$, and gas residence time $\tau \approx$ 0.11-0.15 s.

Depending on the feed gas concentration and bed temperature, experimental data were collected during a time interval of 15 minutes under approximate steady-state conditions. Excess steam was fed in every run for avoiding mass transfer limitations and was condensed and collected downstream. Water consumption was determined by weighing fed water upstream and condensed water downstream and further verified by oxygen mass balance of gases measured by GC. The experiment was terminated by simultaneously shutting down the solar simulator and the steam flow and by cooling the setup with an argon flow.

The fluidized bed reactor exhibited uniform gas-solid contacting surface over the reaction domain, provided efficient heat and mass transport, and served as a suitable tool for measuring interfacial reaction kinetics [49]. Because of the strong mixing conditions, it was reasonable to assume that all fluidized particles were uniformly irradiated and at uniform temperature. Further, since the experimental runs were conducted under high dilution of reactants, the assumption of monolayer chemisorption necessary for the application of the Langmuir-Hinshelwood model, elaborated in Chapter 4.3, was adequate as well.

6.1.3 Results

The measured temperature and off-gas composition of a representative experimental run is shown in Figure 6.4 and the corresponding water and petcoke conversions are shown in Figure 6.5. While the petcoke particles remained in the fluidized bed until they were consumed, typical residence times of the continuous steam flow in the fluidized bed ranged between 0.11 - 0.15 seconds.

The water conversion, X_{H_2O}, is calculated from the oxygen mass balance using the product gas flows $\dot{n}_{CO}$, $\dot{n}_{CO_2}$ (calculated from GC measurements and feed gas flows), and $\dot{n}_{H_2O,0}$ (measured by the water pump), as:

$$X_{H_2O} = \frac{\dot{n}_{CO} + 2\dot{n}_{CO_2}}{\dot{n}_{H_2O,0}} \tag{6.1}$$

The carbon conversion, X_C, is calculated from the carbon mass balance as:

$$X_C = \frac{\int_0^t \dot{n}_{CO} + \dot{n}_{CO_2} + \dot{n}_{CH_4} dt}{n_{C,0}} \tag{6.2}$$

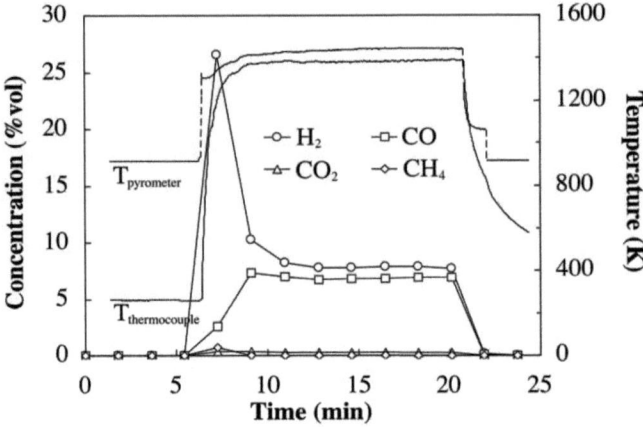

Figure 6.4: Fluidized bed temperature and product gas composition during a representative experimental run with Flexicoke and 10% H_2O-Ar.

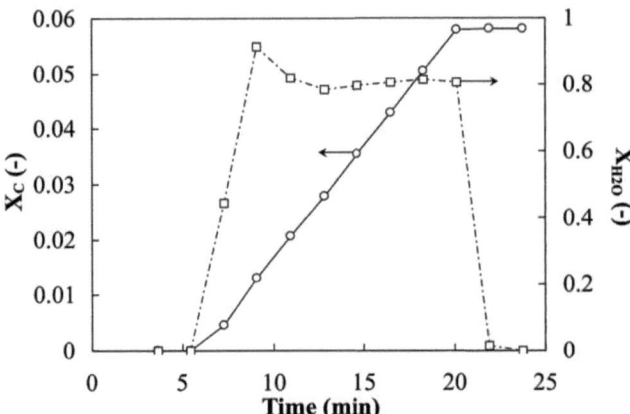

Figure 6.5: Steam and carbon conversion for the experimental run of Figure 6.4 with Flexicoke and 10% H_2O-Ar.

6.1. DFB REACTOR

where $n_{C,0}$ denotes the initial amount of carbon in the bed.

The variation of the product gas composition as a function of the fluidized bed temperature using a feed gas of 10% H_2O-Ar is shown in Figures 6.6 and 6.7 for the gasification of Flexicoke and PD coke, respectively. Main product gases are H_2, CO, and CO_2. CH_4 concentration was less than 0.02%. For Flexicoke, CO_2 concentration reaches a peak of 1.15% at about 1250 K, and becomes negligible at temperatures above 1450 K. For PD coke, the peak of CO_2 is obtained between 1350 and 1450 K.

Complete avoidance of CO_2 presumably takes place above 1700 K. For PD coke, an irregular trend of the off gas composition is observed in the 1350-1450 K interval of Figure 6.7. Heat treatment may diminish reactivity because of a reduction of active sites due to the loss of hydrogen and oxygen atoms in the surface, and because of a reduction in the micro-porosity and carbon edges (the latter serves as active sites like oxygen and hydrogen atoms) via cluster reorganization by thermal annealing [47].

Figure 6.8 shows the gas product relative composition as a function of temperature for the 2 types of petcoke (left: Flexicoke, right: PD coke). Argon and excess steam were omitted. The quality of the syngas produced is described by the H_2/CO and CO_2/CO molar ratios. The H_2/CO molar ratio can be adjusted via the exothermic water-gas shift reaction to meet the requirements for the post-processing of syngas to hydrogen, ammonia, methanol, or Fischer-Tropsch chemicals. The CO_2/CO molar ratio is a measure of the contamination and should be kept preferable as low as possible.

The experimental results of Figure 6.8 are consistent with the thermodynamic predictions of Figures 3.1 and 3.2. Syngas containing approximately an equimolar mixture of H_2 and CO and with a relative CO_2 content of less than 5% was produced at above 1350 K and 1550 K for Flexicoke and PD coke, respectively. The syngas quality is notably higher than the one typically obtained when heat is supplied by internal combustion of petcoke, besides the additional benefit of avoiding contamination by undesirable species. The variation of the H_2S concentration with temperature is shown in Figure 6.9 for both petcokes. As expected, higher H_2S concentrations are found for PD coke because of its higher content of elementary sulfur (see Table 2.2 on page 19).

Pyrolysis

The steam gasification of petcoke is characterized by the concurrent pyrolysis process, namely: the conversion of petcoke to gaseous products in the absence

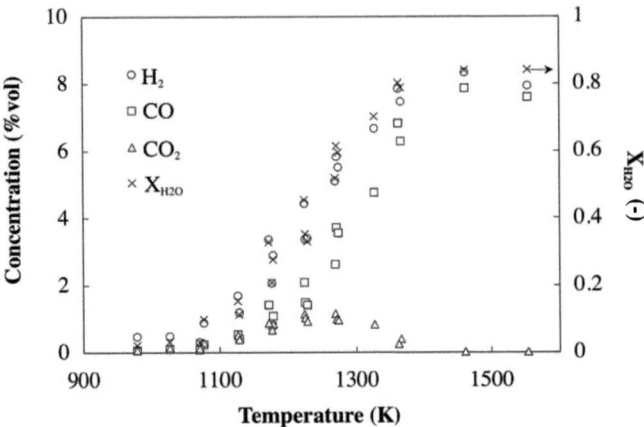

Figure 6.6: Product gas composition and steam conversion as a function of the fluidized bed temperature for Flexicoke gasification using a feed gas of 10% H_2O-Ar.

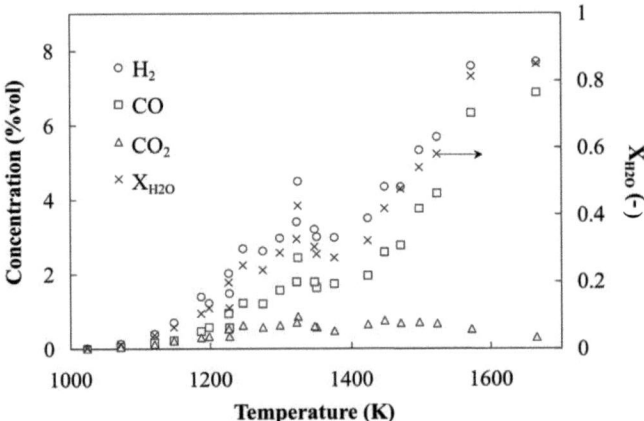

Figure 6.7: Product gas composition and steam conversion as a function of the fluidized bed temperature for PD coke gasification using a feed gas of 10% H_2O-Ar.

6.1. DFB REACTOR

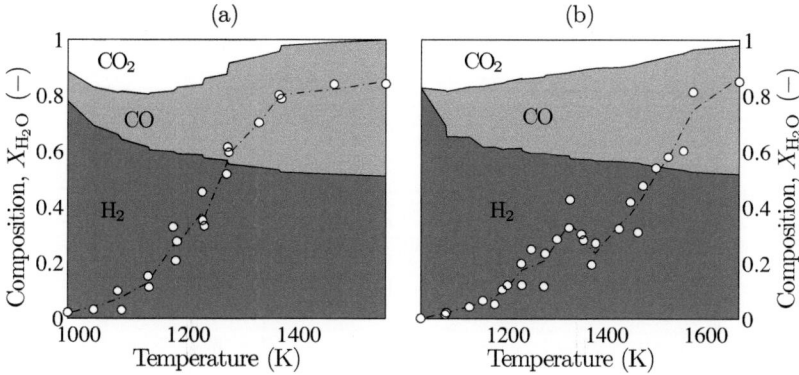

Figure 6.8: Product gas relative composition and steam conversion as a function of temperature for the steam gasification of (a) Flexicoke and (b) PD coke with direct irradiation of the fluidized bed. Argon and excess steam are omitted.

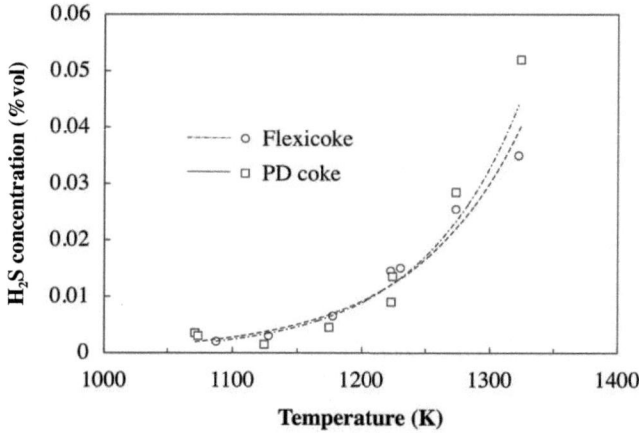

Figure 6.9: Variation of H_2S concentration in the product gas with temperature for the steam gasification of both petcokes.

Table 6.2: Experimental parameters of hydrogen and oxygen released by pyrolysis during *dry* experimental runs.

	Flexicoke	PD coke
Fluidized bed temperature	1390 K	1420 K
Initial sample weight	7.544 g	3.516 g
H-weight fraction	0.0067	0.00414
O-weight fraction	0.00915	0.00146
n_{H_2}	0.0085 mol	0.000685 mol
n_{CO}	0.0018 mol	0.000685 mol
Fraction of H released	34%	75%
Fraction of O released	41%	21%

of an oxidizing agent. In experiments performed with high heating rates ($\approx$ 50 K/s), the rate of pyrolysis-related reactions appears to be significantly higher than that of H_2O gasification-related reactions, as recorded by GC (see for example, Figures 6.4 and 6.5). Immediately after the turning on of the radiation source, but before the steam flow was injected into the reactor, sharp peaks of H_2 and CO - attributed to pyrolysis and desorption processes - were observed in concentrations higher than those obtained during the reaction with steam.

Quantitative analysis of the pyrolysis gas production was determined in *dry* experiments (100% Ar). Figure 6.10 shows the measured reactor temperature and product gas concentration during the pyrolysis of Flexicoke. Integration of the GC peaks over time yielded the fractions of hydrogen and oxygen released during pyrolysis, listed in Table 6.2. 34% and 75% of the hydrogen and 41% and 21% of the oxygen initially contained in the coke were released during pyrolysis of Flexicoke and PD coke, respectively.

The pyrolysis of PD coke was further investigated as a function of temperature: Figure 6.11 presents the variation of the elementary composition of the residual coke with temperature. Clearly recognizable is the progressive carbonization with increasing bed temperature, reaching 95.6% at 1665 K. Hydrogen and oxygen content approached zero at above 1300 K.

Figures 6.12 and 6.13 show the specific surface, the micropore area and volume, and the average pore diameter of PD coke. Vis-à-vis the raw PD coke having a BET specific surface 1 m^2/g, an increase of specific surface was found for all samples subjected to the combined pyrolysis and steam gasification in the quartz fluidized bed reactor. The highest specific surface ($\approx$ 80 m^2/g)

6.1. DFB REACTOR

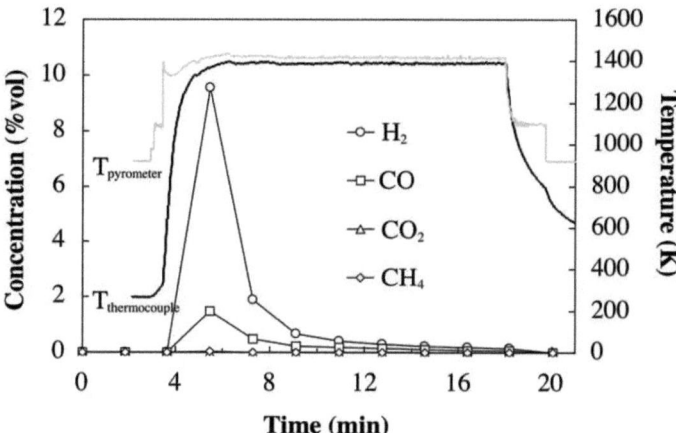

Figure 6.10: Reactor temperature and product gas concentration during a *dry* experimental run with Flexicoke.

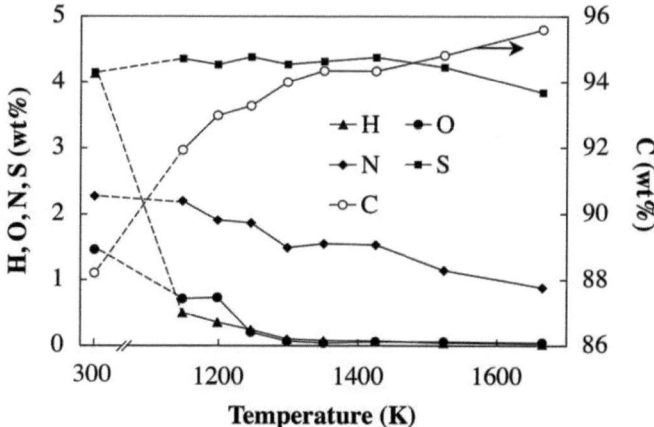

Figure 6.11: Variation of the elementary composition of PD coke with temperature.

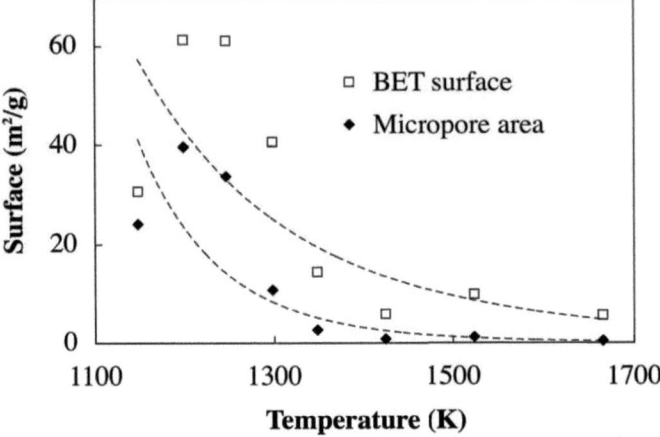

Figure 6.12: Variation of the specific surface and micropore area of PD coke with temperature.

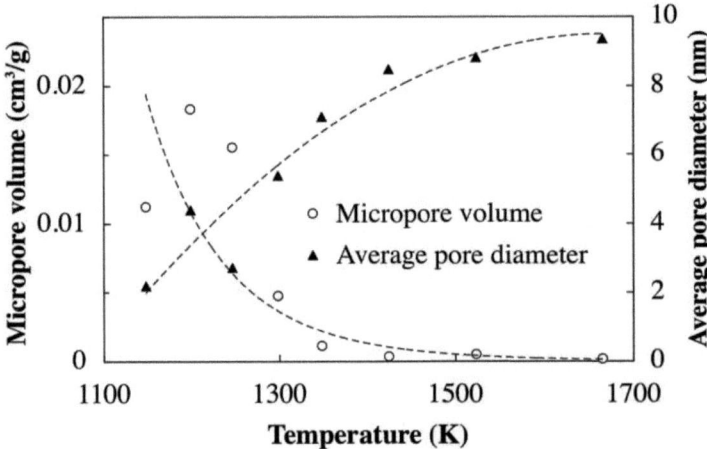

Figure 6.13: Variation of the micropore volume and the average pore diameter of PD coke with temperature.

6.2. IFB REACTOR

as well as micropore area (≈ 40 m^2/g) and volume (≈ 0.02 cm^3/g) are observed for samples treated at temperatures below 1300 K. Higher reactor temperatures result in a dramatic loss of specific surface and the micropore parameters indicate an even stronger decrease of microporosity. At the same time, the average pore diameter is increased by a factor of five from 2 to 10 nm. The carbon conversion of all samples was below 20%. The results correspond to DiPanfilo and Egiebor [18], who found that the surface area of synthetic crude coke is generally increased by activation with steam at 1123 K and report values in the order of 100 m^2/g for samples with a burn-off (carbon conversion) of 20%.

6.2 Fluidized bed reactor with indirect irradiation

6.2.1 Reactor setup

For the steam gasification experiments using indirect irradiation a new reactor setup was designed based on the experiences with the quartz fluidized bed reactor (QFBR) from the previous chapter. The new design is an optimization with regard to mechanical stability and experimental flexibility. In opposition to the DFB setup from the previous chapter, the IFB setup allows to use any kind of fluidization tube with an outer diameter of 2.5 cm. The reactor setup consisted of a water-cooled aluminum plate acting as radiation shield and base plate to mount the open fluidization tube and the surrounding quartz dome separating the reactive atmosphere from ambient air.

Figure 6.14 shows a technical drawing of the reactor setup. The fluidization tube has an outer diameter of 2.5 cm and is 20 cm long. At 9 cm from the bottom end it contains a porous quartz disk acting as bed support. The disk features a lead-through for the bed thermocouple in the form of a pinhole sealed with high temperature cement. The fluidization tube is mounted with o-rings in the water-cooled radiation shield and can be fixed at an arbitrary height. The quartz dome mounted on top of the plate separates the fluidization tube with the reactive gas environment from the ambient. Both the quartz dome and the fluidization tube are sealed against the base plate with o-rings fixed with a sealing unit from the bottom side. The annulus between the quartz dome and the fluidization tube is purged with a sweep argon flow to avoid dead volumes. All gases including the fluidization gas with the gasification products leave the reactor through a horizontal quartz

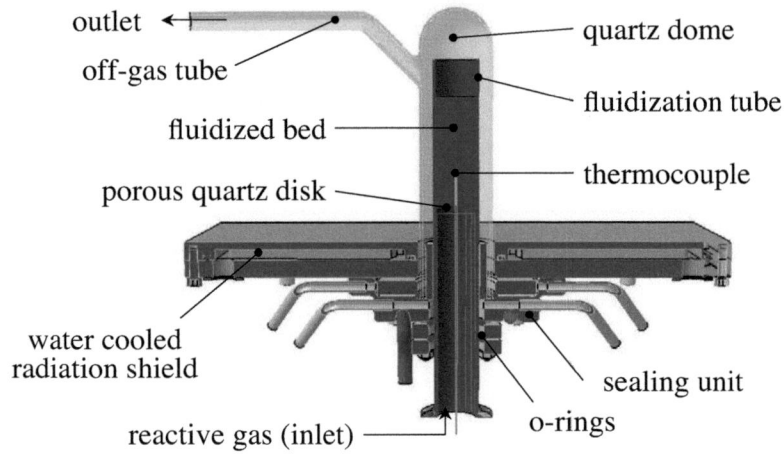

Figure 6.14: Fluidized bed reactor setup for the use of opaque ceramics fluidization tubes with radiation shield and quartz dome.

Table 6.3: Characteristics of the fluidization tubes tested with the setup in Figure 6.14.

material	inner ⌀ (mm)	d_{wall} (mm)	$k^\dagger$ (W/m/K)	$\epsilon^\dagger$ (-)
Al_2O_3	22	1.5	5	0.21
Al_2O_3	20	2.5	5	0.21
SiC	18	3.5	35	0.94
quartz	22	1.5		

† at 1000 °C.

6.2. IFB REACTOR

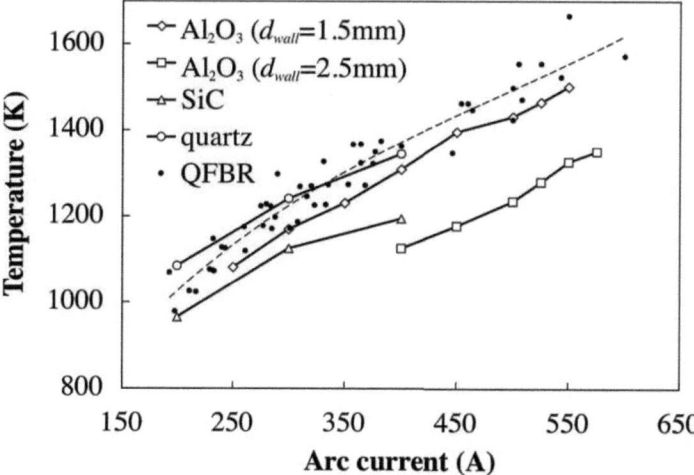

Figure 6.15: Bed temperature as a function of the arc current for different fluidization tubes. Bullets ('•') mark the experiments with the QFBR.

tube located at the dome's top end to the cooling trap and the gas analysis system (GC). The flowchart of the IFB setup is the same as for the previous campaign and is shown in Figure 6.1 on page 86.

Fluidization tubes of different materials with different wall thicknesses and thermal conductivities were tested and compared to a quartz tube representing the radiative heat transfer characteristics of the quartz reactor from the DFB experiments. Table 6.3 presents an overview of the tested fluidization tubes with the respective material properties.

Figure 6.15 shows the dependence of the bed temperature from the arc current for different fluidization tubes. The dots with the polynomial trend line represent the data of the DFB experiments. The quartz fluidization tube of the new setup exhibits the same heat transfer characteristics as the QFBR. The Al_2O_3 tubes and the SiC tubes require higher arc currents associated with higher electrical power input to obtain the same temperatures as the quartz setups.

The IFB experiments presented in Chapter 6.2.3 are performed with the 1.5 mm thick Al_2O_3 tube using a ceramic plug to cover the top and shield the particles in the fluidized bed from direct thermal radiation emitted by the argon arc. In contrast to the QFBR where radiation directly passes through the quartz wall to the coke bed, the fluidized bed of the IFB reactor is fully protected against direct radiation emitted by the argon arc, and the heat transfer is characterized by an additional conduction step across the wall of the opaque fluidization tube.

6.2.2 Experimental procedure

The applied experimental procedure is essentially the same as described in Chapter 6.1.2 for the DFB experiments. A batch of 4 g solid reactants is first heated to the desired reaction temperature under an argon flow and then subjected to continuous H_2O-Ar flow under isothermal conditions. Bed temperatures in the 1000-1600 K range are measured with a type S thermocouple submerged in the fluidized bed. The fluidized bed characteristics are the same as with the QFBR reactor. Depending on the feed gas concentration and bed temperature, experimental data are collected during a time interval of 15 minutes under approximate steady-state conditions. The experiment is terminated by simultaneously shutting down the solar simulator and the steam flow and by cooling the setup with an argon flow.

6.2.3 Results

Design improvements achieved with the new setup allowed to run longer experiments at elevated temperatures and therefore to obtain steady states at different temperatures in ascending series. Figure 6.16 shows temperature and off-gas composition of a representative experimental run with PD coke in the opaque fluidized bed reactor. Figure 6.17 shows the corresponding carbon and steam conversion calculated with Equations (6.1) and (6.2). Pyrolysis of the coke occurs during the first 15 minutes of the experiment used to heat the fluidized bed to the desired temperature.

Gaseous pyrolysis products detected by the GC are mainly methane and hydrogen as well as traces of carbon monoxide and dioxide. The amount of gas released during the pyrolysis of PD coke is notably higher than that previously reported for Flexicoke (cf. Figures 6.4 and 6.5). The release of methane observed in Figure 6.16 affects the carbon inventory of the solid and therefore results in a 7.1% increase of carbon conversion presented in Figure 6.17. Further, the slope of X_C in Figure 6.17 shows that the pyrolysis reaction is significantly faster than the steam gasification reaction.

6.2. IFB REACTOR

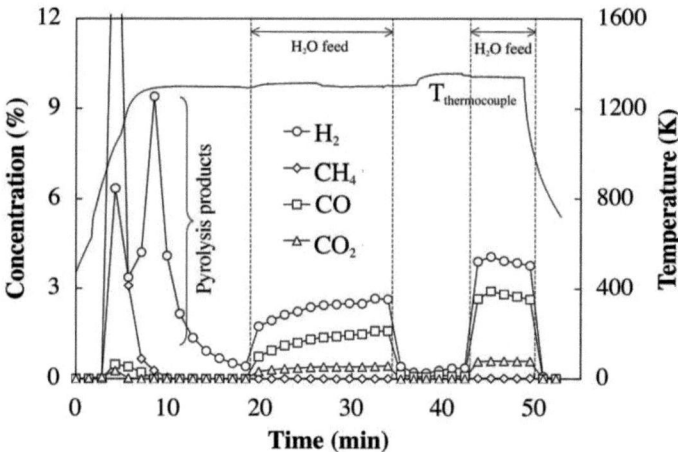

Figure 6.16: Fluidized bed temperature and product gas composition during a representative experimental run with PD coke in 10% H_2O-Ar.

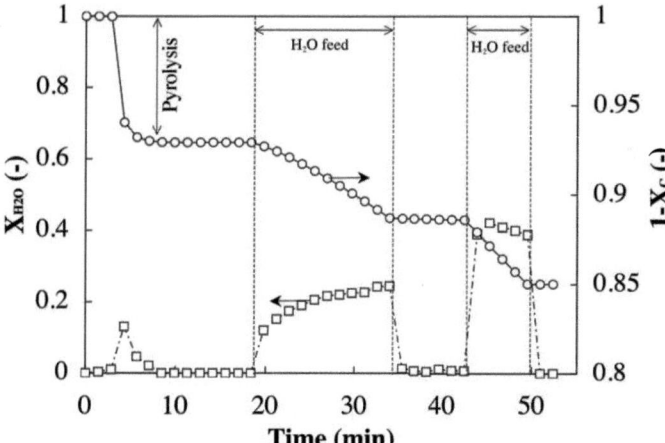

Figure 6.17: Steam and carbon conversion for the experimental run of Figure 6.16 with PD coke and 10% H_2O-Ar.

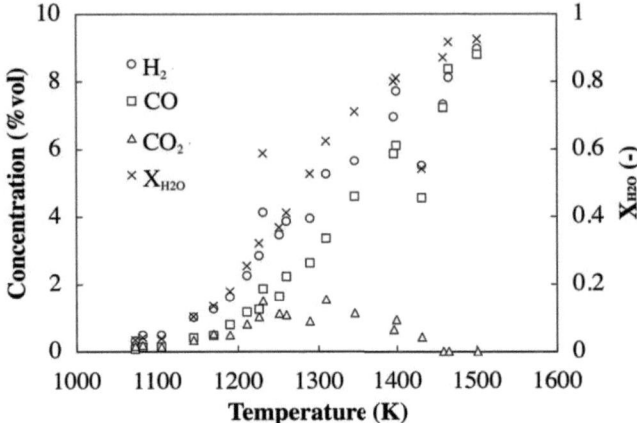

Figure 6.18: Product gas composition and steam conversion as a function of the fluidized bed temperature for the gasification of Flexicoke using 10% H_2O-Ar.

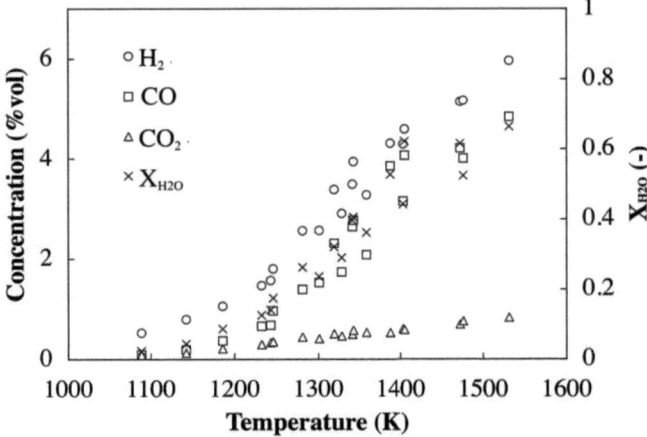

Figure 6.19: Product gas composition and steam conversion as a function of the fluidized bed temperature for the gasification of PD coke using 10% H_2O-Ar.

6.3. KINETIC ANALYSIS

To calculate the rate of gas production under approximate steady state conditions, a time interval comprising at least 5 GC measurements with a standard deviation of less than 10% was selected and the arithmetic mean is used as the characteristic concentration of the interval. Figure 6.18 and 6.19 show the product gas concentration using a 10% H_2O-Ar feed gas as a function of the fluidized bed temperature for Flexicoke and PD coke, respectively. Main products are hydrogen, carbon monoxide and carbon dioxide at concentrations comparable to previous experiments (cf. Figures 6.6 and 6.7).

Steam gasification of Flexicoke shows a carbon dioxide maximum close to 2% at 1300 K and produces pure syngas at higher temperatures, whereas gasification of PD coke resulted in a constant increase of carbon dioxide concentration over the whole experimental temperature range, against the findings of the previous campaign. Methane appeared in considerable amounts during the pyrolysis of PD coke but only at a very low level during gasification with steam ($y_{CH_4} \leq 0.020$ for Flexicoke and $y_{CH_4} \leq 0.025$ for PD coke) and is therefore omitted from the plots.

Although not as evident as in the experiments with direct irradiation of PD coke, a decrease of reactivity at temperatures around 1400 K is found in concentrations and steam conversion shown in Figure 6.19. Compared to the experiments with direct irradiation, a slight shift of the deactivation towards higher temperatures can be observed.

Figure 6.20 shows the product gas relative composition and steam conversion as a function of temperature for the steam gasification of (a) Flexicoke, and (b) PD coke using indirect irradiation of the fluidized bed. Argon and excess steam are omitted. As previously mentioned, essentially pure syngas is obtained by steam gasification of Flexicoke at temperatures above 1460 K, whereas carbon dioxide concentration did not decrease in the investigated temperature interval.

6.3 Kinetic analysis

The steam gasification reaction affects the concentration of several gases including H_2O, H_2, CO_2, and CO. The intrinsic reaction rates are expressed in terms of moles of gas per unit coke surface and time

$$r_i = \frac{1}{m_{coke}\, a} \frac{dn_i}{dt} \tag{6.3}$$

where m_{coke} is the weight of the coke sample obtained from an integral carbon mass balance, a is the specific active surface in (m^2/g) (cf. Table 6.1), and

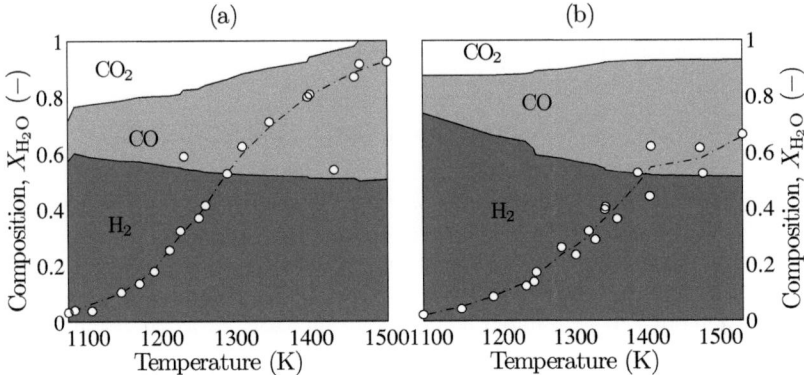

Figure 6.20: Product gas relative composition and steam conversion as a function of temperature for the steam gasification of (a) Flexicoke, and (b) PD coke with indirect irradiation of the fluidized bed. Argon and excess steam are omitted.

$\dot{n}_i = dn_i/dt$ are the gas flows calculated from the mass flow rates and the GC measurements.

The rate laws used to model the chemical reactions pertinent to the reactive gasification are based on the extended mechanism derived in Chapter 4.3. Due to limitations of the experimental setup, all runs were performed with a dilute reactive gas consisting of 10% H_2O in Ar to avoid steam condensation in the feed lines. The restriction to low partial pressures allows using a simplified version of the Langmuir-Hinshelwood rate laws (4.77) to (4.80): For $p_i \to 0$, the limit of the denominator in Equations (4.77) to (4.80) is 1. The inhibition terms in the denominator can thus be neglected leading to the following set of first order rate laws:

$$r_{H_2O} = -K_2 p_{H_2O} - K_2 K_3 p_{H_2O} p_{CO} \qquad (6.4)$$
$$r_{H_2} = -r_{H_2O} \qquad (6.5)$$
$$r_{CO_2} = -K_1 p_{CO_2} + K_2 K_3 p_{H_2O} p_{CO} \qquad (6.6)$$
$$r_{CO} = 2K_1 p_{CO_2} + K_2 p_{H_2O} - K_2 K_3 p_{H_2O} p_{CO} \qquad (6.7)$$

The carbon consumption rate of the steam gasification reaction is obtained

6.3. KINETIC ANALYSIS

from a carbon mass balance:

$$-r_C = r_{CO} + r_{CO_2} \tag{6.8}$$
$$= K_1 p_{CO_2} + K_2 p_{H_2O} \tag{6.9}$$

The new Equations (6.4) to (6.7) are linearizations of the Langmuir-Hinshelwood rate laws (4.77) to (4.80) that apply to experiments with low partial pressures only. The use of the linearized system can be justified by evaluating the inhibition term with the rate data obtained from the experiments in the thermobalance. Evaluation of the inhibition term assuming a reactive gas with 10% H_2O-Ar and using the Arrhenius parameters presented in Table 5.7 results in a maximum linearization error of 7.8% for the gasification of PD coke, whereas it is negligible for Flexicoke.

For the kinetic modeling the coke particles are assumed to be isothermal. Müller [63] showed for gasification experiments with activated charcoal particles bigger than the coke used for this study that heat transfer resistances in the particle can be neglected in the 1000-1300 K temperature range. In opposition to heat transfer, the kinetic model accounts for mass transfer limitations inside the particle using the effectiveness model defined by Equation (4.15) in Chapter 4.1.1

$$\eta_P = \frac{r_{obs}}{r_{intr}}$$

and the experimental correlation for η_P presented in Chapter 5.2.2.

The calculation of the rate constants from experimental rate data measured at different bed temperatures is performed using a technique developed by [63]: The reactor is considered as a two phase fluidized bed assuming ideal mixing for the solid phase and plug flow for the gas phase. Applying mass balance for each species i over a differential reactor element of particle mass along the reactor axis yields the performance equation for a plug flow reactor (cf. [49])

$$\dot{n}_i = r_i \, dA_{coke} = r_i \, a \, dm_{coke} \tag{6.10}$$
$$\text{with} \quad i = H_2O, H_2, CO_2, CO$$

where $\dot{n}_i$ (mol/s) are the molar flows in the axial direction, and r_i are the rates defined by Equations (6.4) to (6.7). The partial pressures in the rate expressions are calculated assuming ideal gases

$$p_i = P_{tot} \frac{\dot{n}_i}{\sum_j \dot{n}_j} \tag{6.11}$$

The system of four coupled differential Equations (6.10) is solved numerically using a variable order solver for stiff initial value problems, based

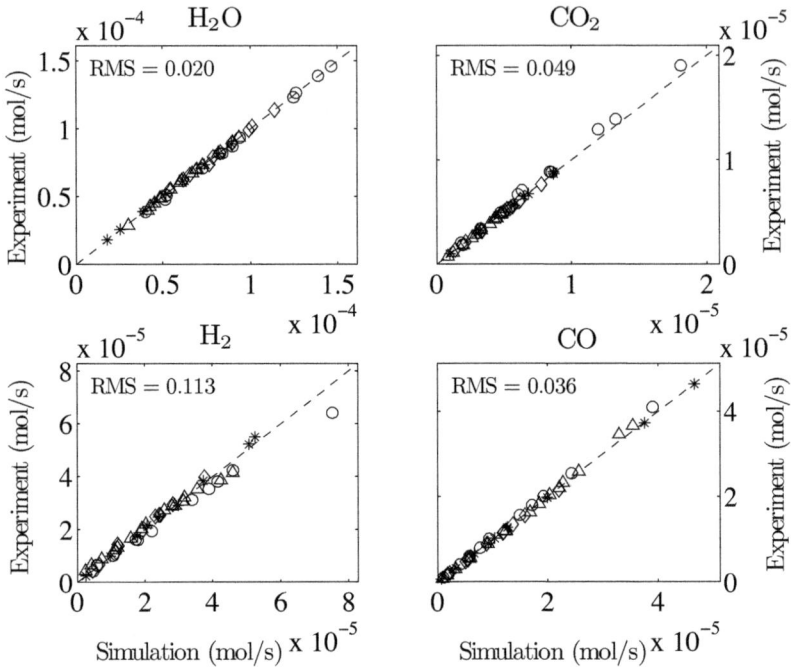

Figure 6.21: Experimentally measured vs. modeled product flow rates ('∗' Flexicoke DFB, 'o' Flexicoke IFB, '◇' PD coke DFB, '△' PD coke IFB).

on the numerical differentiation formulas (MATLAB [37] ode15s). The Arrhenius parameters associated with the kinetic constants K_i are calculated iteratively by minimizing the difference between the theoretically calculated and experimentally measured molar flow rates of reactants and products at the reactor exit using the Nelder-Mead Simplex Method for multidimensional unconstrained nonlinear minimization (MATLAB [37] fminsearch).

Figure 6.21 shows the theoretically modeled versus the experimentally measured product flow rates. The dashed line represents perfect match. The dimensionless root mean square (RMS) relative error is 0.020 for H_2O, 0.049 for CO_2, 0.113 for H_2 and 0.036 for CO.

Figure 6.22 shows an example of a calculated concentration profile in

6.3. KINETIC ANALYSIS

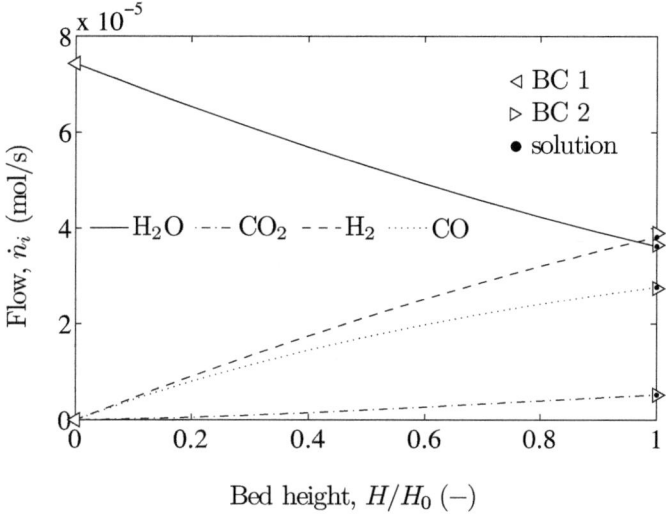

Figure 6.22: Concentration profile along the dimensionless reactor axis and boundary conditions (◁ and ▷) for the steam gasification of PD coke in 10% H_2O-Ar at 1572 K.

the direction of the reactor axis for a gasification experiment with PD coke in the DFB reactor. The measured bed temperature is 1572 K and the fluidization gas is 10% H_2O-Ar. Further, the x-axis is the dimensionless bed height, H/H_0, and the gas composition is reported in terms of the molar flow rates of H_2O, H_2, CO_2, and CO. The triangles at the beginning and the end of the reactor mark the boundary conditions given by the experiment. The calculated solution of Equation (6.10) is indicated by lines across the bed region and bullets at $H/H_0 = 1$ for H_2O, H_2, CO_2, and CO. Right at the reactor inlet, steam starts to react with the coke in the fluidized bed producing H_2 and CO. The rate of CO_2 formation is zero at the reactor inlet since it is a secondary product requiring the presence of CO.

The pre-exponential factor, k_0, and the apparent activation energy, E_A, associated with the rate constants K_1a, K_2a, and K_3 according to Equation (4.36) are obtained from an Arrhenius analysis. The respective plots are shown in Figures 6.23, 6.24 and 6.25 for the DFB and IFB experiments, each including a data set for Flexicoke and PD coke, respectively. E_A and k_0 are

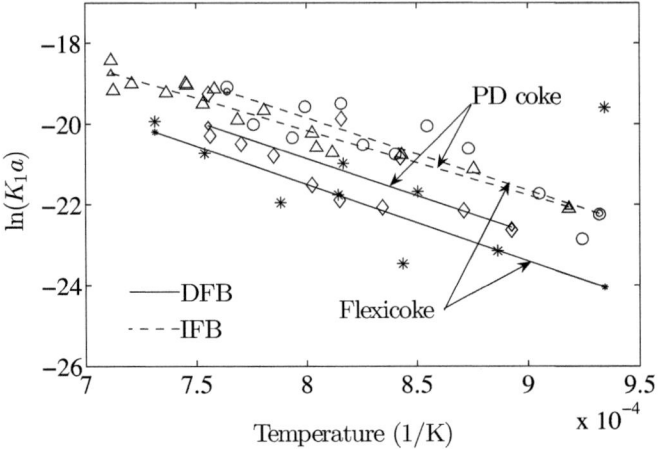

Figure 6.23: Arrhenius plots of $K_1 a$.
∗ Flexicoke DFB, ○ Flexicoke IFB, ◇ PD coke DFB, △ PD coke IFB.

obtained by robust linear regression using an iteratively re-weighted least squares algorithm. Numerical values are presented in Table 6.4 for Flexicoke and in Table 6.5 for PD coke, respectively.

Since the K_i represent complex reaction mechanisms rather than elementary steps, a negative value for the apparent activation energy, as in the case for K_3, does not necessarily imply inconsistency with the transition state theory. Following the definition of K_3 in Equation (4.81), it rather implies that the activation energy for the rate of the elementary reaction of adsorbed oxygen with carbon to form CO (Equation 4.56) is higher than the one for the rate of the elementary reaction of absorbed oxygen with CO to form CO_2 (Equation 4.54) [63].

The error of the rate constants K_i in Tables 6.4 and 6.5 is reported in terms of the root mean square (RMS) relative error between experimental data and regression and the coefficient of determination, R^2 (cf. Equation (5.6) on page 72). The highest accuracy is found for $K_2 a$ exhibiting small relative errors between 1.0 and 1.6% and determination coefficients between 0.90 and 0.96. $K_2 a$ relates to the formation of adsorbed hydrogen from steam and mainly controls the conversion of steam to hydrogen in the linear system (6.4) to (6.7).

6.3. KINETIC ANALYSIS

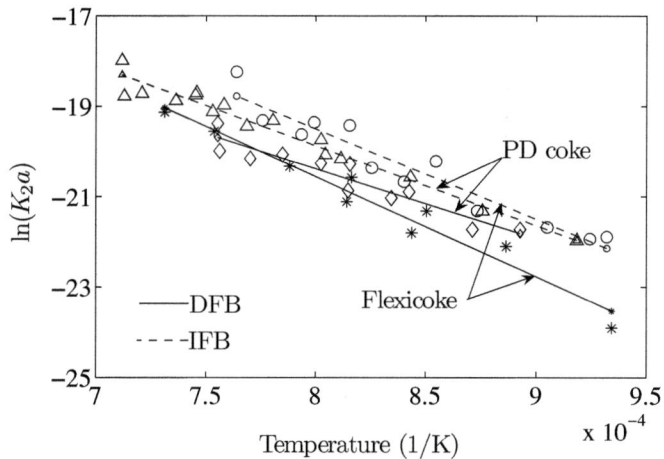

Figure 6.24: Arrhenius plots of K_2a.
∗ Flexicoke DFB, ○ Flexicoke IFB, ◇ PD coke DFB, △ PD coke IFB.

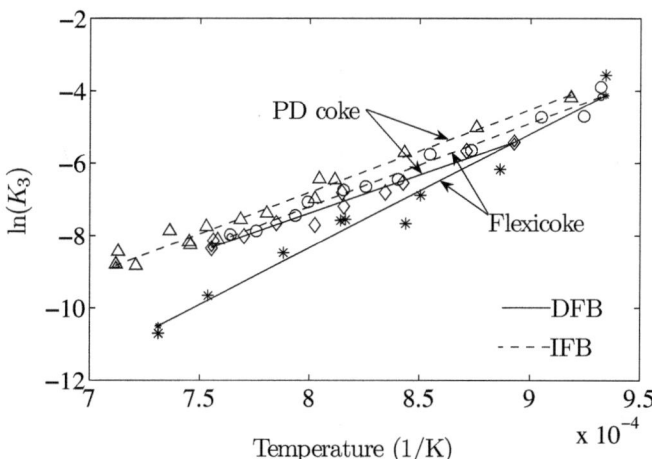

Figure 6.25: Arrhenius plots of K_3.
∗ Flexicoke DFB, ○ Flexicoke IFB, ◇ PD coke DFB, △ PD coke IFB.

Table 6.4: Arrhenius parameters for the rate law of steam gasification of Flexicoke and PD coke respectively, measured with the DFB reactor (Chapter 6.1).

Rate constant		k_0 [K_i]	E_A (kJ/mol)	RMS (−)	R^2 (−)
Flexicoke direct (1070-1367K)					
$K_1 a$	(mol/g/s/Pa)	$1.792 \cdot 10^{-3}$	157.7	$8.02 \cdot 10^{-2}$	0.033
$K_2 a$	(mol/g/s/Pa)	$6.093 \cdot 10^{-2}$	184.5	$1.25 \cdot 10^{-2}$	0.960
K_3	(1/Pa)	$2.879 \cdot 10^{-15}$	−261.3	$7.06 \cdot 10^{-2}$	0.959
PD coke direct (1120-1323K)					
$K_1 a$	(mol/g/s/Pa)	$2.227 \cdot 10^{-3}$	153.3	$3.06 \cdot 10^{-2}$	0.618
$K_2 a$	(mol/g/s/Pa)	$3.387 \cdot 10^{-4}$	128.8	$1.10 \cdot 10^{-2}$	0.900
K_3	(1/Pa)	$2.705 \cdot 10^{-11}$	−176.1	$2.47 \cdot 10^{-2}$	0.965

Table 6.5: Arrhenius parameters for the rate law of steam gasification of Flexicoke and PD coke respectively, measured with the IFB reactor (Chapter 6.2).

Rate constant		k_0 [K_i]	E_A (kJ/mol)	RMS (−)	R^2 (−)
Flexicoke indirect (1073-1309K)					
$K_1 a$	(mol/g/s/Pa)	$4.220 \cdot 10^{-3}$	149.5	$2.35 \cdot 10^{-2}$	0.805
$K_2 a$	(mol/g/s/Pa)	$3.062 \cdot 10^{-2}$	166.4	$1.61 \cdot 10^{-2}$	0.923
K_3	(1/Pa)	$7.253 \cdot 10^{-12}$	−191.7	$3.47 \cdot 10^{-2}$	0.982
PD coke indirect (1089-1405K)					
$K_1 a$	(mol/g/s/Pa)	$6.667 \cdot 10^{-4}$	133.5	$1.27 \cdot 10^{-2}$	0.932
$K_2 a$	(mol/g/s/Pa)	$3.358 \cdot 10^{-3}$	147.3	$1.03 \cdot 10^{-2}$	0.964
K_3	(1/Pa)	$1.245 \cdot 10^{-11}$	−190.2	$2.76 \cdot 10^{-2}$	0.975

6.3. KINETIC ANALYSIS

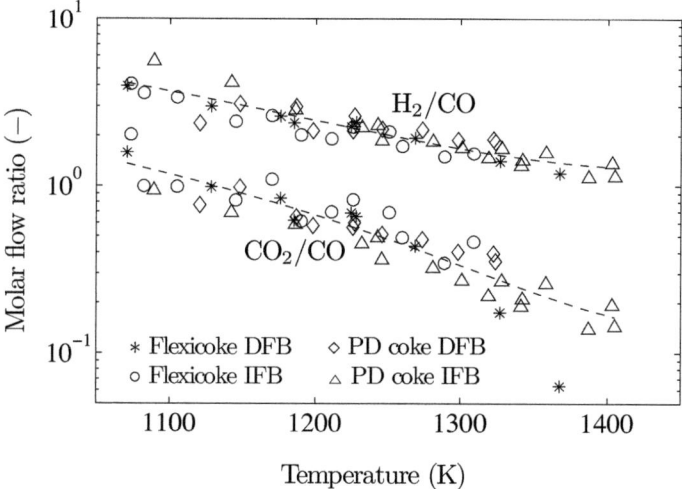

Figure 6.26: Temperature dependence of the H_2/CO and CO_2/CO molar ratio in the product gas for the experiments with the DFB and IFB reactor.

The relative errors of K_1a and K_3 are bigger. K_1 is the rate constant associated with the dry gasification, a reaction that is not measured directly because the feed gas contains no CO_2. Further, low determination coefficients indicate a poor quality of the linear regression for K_1, especially for the experiments with direct irradiation of the fluidized bed. K_3 appears only as a product with K_2 and is related with the CO shift [cf. Equation (3.6) on page 24 and Equation (6.12)]. The calculated values for K_3 are very small: The pre-exponential factor associated with K_3 is about ten orders of magnitude smaller and the activation energy is the largest of all. Tests with K_3 set to zero showed that the overall outcome of the kinetic model is hardly affected by this rate constant.

Figure 6.26 shows the H_2/CO and CO_2/CO molar ratios considered as a quality indicator for the produced syngas [63]. In accordance with the thermodynamic considerations in Chapter 3.1, the H_2/CO decreases with increasing temperature and asymptotically reaches a value in the order of one, depending on the stoichiometry of the gasified coke, while the CO_2/CO ratio goes to zero. Figure 6.26 further shows that there is no difference with respect to the temperature dependence of the off-gas composition for both the two coke types and the DFB and IFB experiments. The composition of the produced gas depends only on the temperature, and the polynomial trend-line anticipates essentially pure syngas above 1500 K.

The product composition can be further adjusted in a water-gas shift reactor according to

$$H_2O + CO \xrightarrow{k_{wgs}} CO_2 + H_2 \qquad (6.12)$$

to meet the requirements for the post processing of syngas to hydrogen, ammonia, methanol or Fischer-Tropsch chemicals.

6.3.1 Comparison of the DFB and IFB rate data

In order to enable a quantitative comparison of the DFB and IFB gasification kinetics, the carbon consumption rate, $-r_C$, is calculated for PD coke and Flexicoke using Equation (6.9) and the rate data presented in Tables 6.4 (DFB) and 6.5 (IFB). Figure 6.27 shows the respective $-r_C$ in the 900-2000 K temperature interval for Flexicoke (a) and PD coke (b) under the assumption of a reactive gas that contains 10% steam and 5% carbon dioxide in argon. The dashed line further marks the corresponding carbon consumption rate obtained from the TG experiments in Chapter 5, serving as a reference.

As long as the low temperature range is considered, the carbon consumption rate, $-r_C$, shows reasonable consistency among the IFB, DFB, and TG experimental data for both Flexicoke and PD coke. At temperatures exceeding the interpolation domain ranging up to 1400 (DFB/IFB) and 1300 K (TG), respectively, it is observed that the DFB and IFB rates fall behind those measured with the TG. This observation is explained by the difficulties associated with the temperature measurement in the fluidized bed. Although the coke particles in the bed shield the thermocouple from thermal radiation, a certain amount of radiation coming from the argon arc passes the fluidized bed because of its high void fraction and hits the thermocouple.

The absorption of radiative energy is proportional to the fourth power of temperature according to

$$q'' = \epsilon \sigma T^4 \qquad (6.13)$$

while the convective heat transfer is linearly depending from the temperature dependence to the surrounding gas phase,

$$q'' = h \cdot \Delta T \qquad (6.14)$$

Thus, the radiative contribution gets more important at higher temperatures letting the error increase. The relative location of the IFB and DFB rate is

6.3. KINETIC ANALYSIS

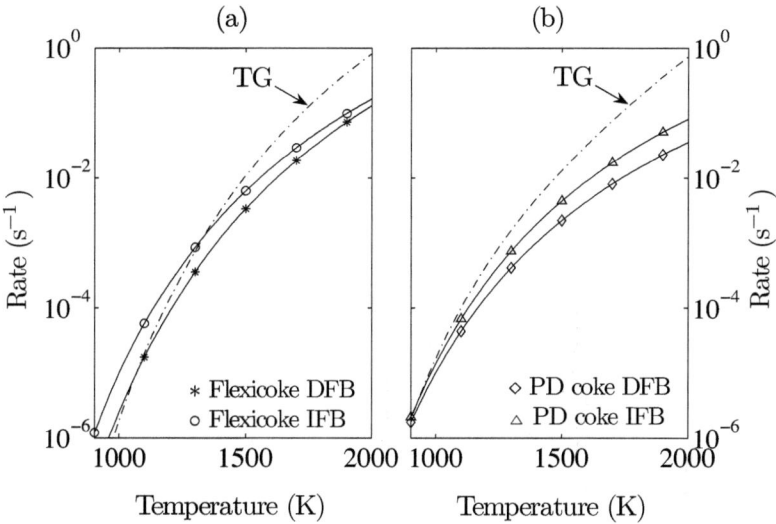

Figure 6.27: Comparison of the carbon consumption rate, $-r_C$, assuming $p_{H_2O} = 0.1$ bar and $p_{CO_2} = 0.05$ bar for the steam gasification of Flexicoke (a) and PD coke (b) using the TG as well as the DFB and IFB setups.

explained with the same arguments because the opaque fluidization tube acts as a radiation shield, which reemits part of the incident radiation to the fluidized bed and the surroundings.

Based on the above conclusions it is found that the slight differences between the DFB and IFB rate data observed for Flexicoke and PD coke, respectively, are not the evidence for a different chemistry, but rather of experimental nature, caused by the difficulties related with the exact measurement of the solid temperature. Because TG setup provides the most unaltered temperature measurement, the TG data is assumed to be the most reliable with regard to the temperature of the solid coke.

As Figure 6.26 further shows, the heat transfer mode does not have an effect on the composition of the product gas. Also, it is important to consider the differences between the Arrhenius parameters k_0 and E_A of the DFB and IFB experiments in front of the quality of the linear fit in Figures 6.23 to 6.25. At least part of the difference between the DFB and IFB rate for Flexicoke is caused by the considerable spread in the respective experimental rate data.

Summarizing the aforementioned findings, one can say that the differences observed between the two experimental series using the DFB and the IFB reactor setup are rather the consequence of limitations associated with the experimental procedure than the consequence of a different chemistry. The two setups are thus equivalent from a chemical point of view. Never the less it is important to recognize that the heat transfer of the DFB setup is better than that of the IFB, having an economical impact, as the former generally achieves a better performance with a given solar power input.

6.4 Summary and conclusions

Steam gasification of petroleum coke was investigated with laboratory scale fluidized bed reactors featuring two different allothermic heat transfer modes. The first reactor consisted of a quartz fluidization tube and the particles in the fluidized bed are heated by direct irradiation from the ETH's high-flux solar simulator. The second reactor allowed using different opaque fluidization tubes and heat is transferred by conduction across the tube wall. The effectiveness of the heat transfer from the argon arc to the fluidized bed is quantified in terms of the arc current. As expected, the DFB setup and the IFB setup with the quartz fluidization tubes achieved the best results.

Two different types of Venezuelan petroleum cokes, a Flexicoke and a delayed coke were ground and sieved to particle sizes in the 250-355 μm range and gasified at temperatures between 1000 and 1600 K. Maximum steam conversion in the order of 85-90% was obtained at temperatures above 1500 and 1600 K for Flexicoke and PD coke, respectively, producing a syngas consisting of an equimolar H_2-CO mixture. Maximum CO_2 production in the order of 10-20% of the steam feed was observed at 1200-1300 K for Flexicoke and 1400-1600 K for PD coke. Despite a small temperature difference of the product and conversion maxima, the H_2/CO and CO_2/CO ratio considered as a quality measure for the syngas do neither differ significantly from coke to coke nor from the indirect to the direct irradiated setup. Sulfur contained in the order of 2-4% of the raw coke is converted to mainly H_2S at a rate increasing in parallel to the carbon consumption.

A set of linearized rate laws based on the extended mechanism of Chapter 4.3 was used to model the chemical reaction. The kinetic parameters and their Arrhenius-type temperature dependence are determined from experimental data assuming plug flow for the fluidization gas and solving the mass balance over a differential reactor element for every relevant component of the gas. The temperature interval for the evaluation of the rate data was

6.4. SUMMARY AND CONCLUSIONS

limited to values below 1400 K. At higher temperatures steam conversion was going to completion and the reaction became mass transfer limited. PD coke further showed a significant reactivity loss due to thermal deactivation, preventing the extraction of meaningful kinetic data.

The rate constants and the resulting reaction rates for the steam gasification of Flexicoke and PD coke that were obtained from the experiments with the DFB and the IFB reactor setups, respectively, did not show a significant difference. In fact, the differences found between the two types of petcoke is larger than the differences found between the experiments with direct and indirect irradiation of the fluidized bed.

Chapter 7

SynPet 5 kW process reactor[1]

This chapter elaborates on the design and experimental performance of the 5 kW SynPet reactor, which was used to demonstrate the solar thermal gasification of petcoke at the pilot scale. The reactor was successfully tested at PSI's solar furnace during fall 2004 and served later on as a model for the scale-up to a nominal thermal power of 500 kW. The reactor design and the experiments presented in this chapter will further serve as the experimental boundary for the mathematical reactor model to be presented in Chapter 8.

7.1 Reactor design

Figure 7.1 shows a schematic of the reactor configuration. It consists of a cylindrical cavity-receiver, 210 mm length, 120 mm inside diameter, that contains a 5 cm diameter opening - the aperture - to let in concentrated solar power. The cavity-type geometry is designed to effectively capture the incident solar radiation; its apparent absorptance[2] is estimated to exceed 0.95.

---

[1]The SynPet reactor design and the experimental data obtained during the experimental campaign at PSI's solar furnace [32] are the result of a joint group effort at the Professorship in Renewable Energy carriers at ETH Zurich. Experimental testing of the reactor in the solar furnace was performed during fall 2004 by P. Haueter, A. Z'Graggen, and the author. Material from this chapter has been published in 'A. Z'Graggen, P. Haueter, D. Trommer, et al. Hydrogen production by steam gasification of petroleum coke using concentrated solar power - II. Reactor design, testing and modeling. *International Journal of Hydrogen Energy*, 2006. 31(6): p. 797-811'.

[2]The apparent absorptance is defined as the fraction of energy flux emitted by a blackbody surface stretched across the cavity opening that is absorbed by the cavity walls. For a cylindrical cavity having a ratio of cavity diameter to aperture diameter equal 2 and a ratio of cavity depth to aperture diameter equal 2.7, the apparent absorptance is greater

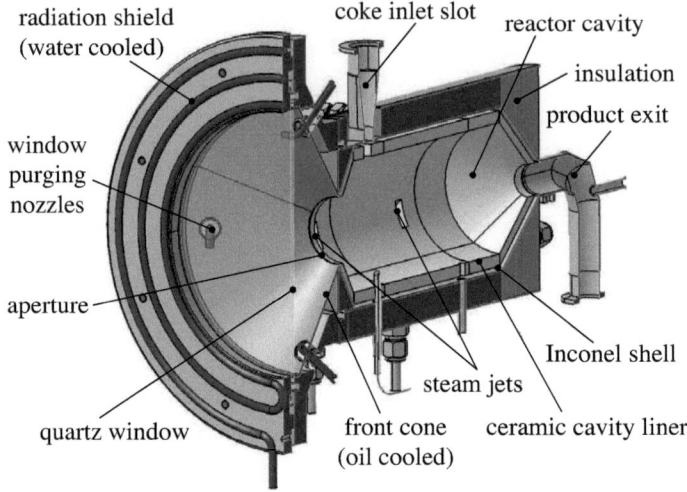

Figure 7.1: Scheme of solar chemical reactor configuration for the steam gasification of petcoke.

The cavity is made out of Inconel-601, lined with Al_2O_3, and insulated with Al_2O_3/ZrO_2 ceramic foam. The aperture is closed by a 0.3 cm thick clear fused quartz window, mounted in a water-cooled aluminum ring that also serves as a shield for spilled radiation. The window is actively cooled and kept clear from particles and/or condensable gases by means of an aerodynamic protection curtain created by a tangential flow through four tangential nozzles combined to a radial flow through a circular gap. In front of the aperture, the cavity-receiver is equipped with a diverging conical funnel for mounting the window 8 cm in front of the focal plane, where the radiation intensity is about 10 times smaller and dust deposition is unlikely to occur. Since radiation spillage can reach flux concentrations ratios[3] greater than 1000 suns, this component is actively oil-cooled and kept in the range 393-453 K to prevent steam condensation.

Steam and particles are injected separately into the reactor cavity permitting individual control of mass flow rates and stoichiometry. Steam is

---

than 0.95 for a surface absorptivity greater than 0.5 [77].

[3]The solar flux concentration ratio is defined as the incident radiative power flux normalized by 1 kW/m^2, and is often reported in units of suns.

introduced through several ports. Based on flow visualization experiments using a plexiglas model and CFD simulations, best flow patterns in terms of residence time and flow stability were obtained with two sets of four symmetrically distributed tangential nozzles, located in planes 3 and 8 cm behind the aperture plane, as shown in Figure 7.1. The petcoke feeding unit is positioned on the top of the reactor vessel with its inlet port located at the same plane as the primary steam injection system, allowing for the immediate entrainment of particles by the steam flow. Inside the cavity, the gas-particle stream forms a vortex flow that progresses towards the rear along a helical path.

With this arrangement, the petcoke particles are directly exposed to the high-flux solar irradiation, providing efficient heat transfer directly to the reaction site. Energy absorbed by the reactants is used to raise their temperature to above about 1300 K and to drive the gasification reaction. Reaction products exit through a 24 mm diameter outlet tube at the rear side of the cavity. This reactor concept has been applied successfully for the solar combined ZnO-reduction and CH_4-reforming [83] and for the solar thermal cracking of CH_4 [35]. These and previous experimental and modeling studies have pointed to the efficient solar energy absorption and heat transfer by the direct solar irradiation of particle suspensions [10, 17, 45, 34].

7.2 Experimental setup

The complete experimental setup is depicted in Figure 7.2. Experimentation was carried out at the PSI's solar furnace [32]. This solar research facility consists of a 120 m^2 sun-tracking heliostat in-axis with an 8.5 m diameter paraboloidal concentrator, and delivers up to 40 kW at peak concentration ratios exceeding 5000 suns. A Venetian blind-type shutter located between the heliostat and the concentrator controls the power input to the reactor. Radiative solar flux intensities were measured optically with a calibrated CCD camera by recording the image of the sun on a water-cooled Al_2O_3-coated Lambertian (diffusely reflecting) plate positioned at the focal plane.

The reactor was positioned with its aperture at the focal plane and intercepting the regions of maximum solar flux intensity. Integration of the incident radiative flux over the reactors aperture yielded the solar power input, $\dot{Q}_{solar}$. The accuracy of the optical measurement combined with the reactors misalignment[4] led to an error of $^{+9}_{-13}$% for $\dot{Q}_{solar}$.

---

[4]Error due to the reactor's misalignment has been estimated by Monte-Carlo ray tracing

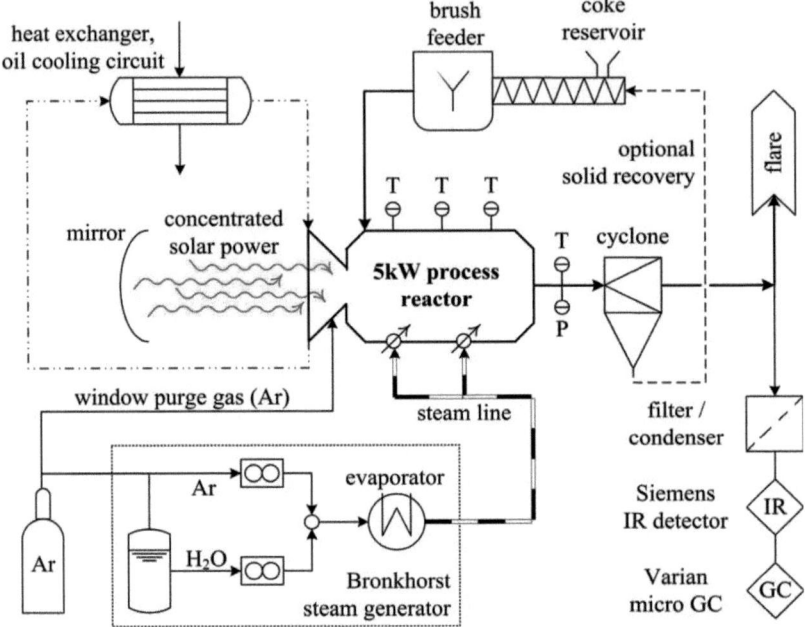

Figure 7.2: Schematic of the experimental setup at PSI's solar furnace.

Reactor wall temperatures were measured in 12 locations with thermocouples type K, inserted in the Inconel walls and not exposed to direct irradiation. The nominal cavity temperature, $T_{reactor}$, was measured with a solar-blind pyrometer that is not affected by the reflected solar irradiation because it measures in a narrow wavelength interval around 1.39 m where solar irradiation is mostly absorbed by the earth atmosphere [91].

Gas flows were controlled using Bronkhorst HI-TEC electronic flow controllers. The coke-feeding ratio was determined by measuring the weight difference of the particle conveyor, with ±5% accuracy. The reactors pressure was monitored with pressure transducers, while a pressure safety valve prevented overpressure derived from a fivefolded volumetric growth due to gas formation and thermal expansion. The gaseous products were analyzed

---

assuming up to 1 cm displacement between the apertures center and the location of peak radiative flux at the focal plane.

on-line by GC (the same instruments was used as for the experiment with the TG setup, cf. Chapter 5.1.1 on page 57). Gas analysis was supplemented by IR-based detectors for CO, CO_2, and CH_4 (Ultramat-23 by Siemens), and a thermal conductivity-based detector for H_2 (Calomat-6 by Siemens). Representative solid product samples collected at the filter downstream of the reactor were examined by scanning electron micrography.

The water conversion, X_{H_2O}, was calculated from the oxygen mass balance using the product gas flows $\dot{n}_{CO}$, $\dot{n}_{CO_2}$ (calculated from GC measurements and feed gas flows) as

$$X_{H_2O} = \frac{\dot{n}_{CO} + 2\dot{n}_{CO_2}}{\dot{n}_{H_2O,0}} \qquad (7.1)$$

where $\dot{n}_{H_2O,0}$ denotes the molar rate of water fed by the pump. The carbon conversion, X_C, is calculated from the carbon mass balance as

$$X_C = \frac{\dot{n}_{CO} + \dot{n}_{CH_4} + \dot{n}_{CO_2}}{\dot{n}_{C,0}} \qquad (7.2)$$

where $\dot{n}_{C,0}$ denotes the molar rate of carbon fed by the calibrated feeder.

7.3 Results

All runs were performed with PD coke particles of 1.2 μm mean diameter obtained by grinding raw PD coke in a jet mill. Reactants were continuously fed at a mass flow rate in the range of 1.85-4.45 g petcoke/min and 3.68-9.04 g H_2O/min. Feeding temperatures were at ambient and 423 K for petcoke and steam, respectively.

The average residence time of the reactants in the reactors cavity was calculated with Equation (7.3) from [102] for a CO_2-free syngas,

$$\tau = \frac{PV}{RT}\left(\dot{n}_{Ar} + \dot{n}_{H_2O} + \dot{n}_C\left(1 + \frac{X_C}{\ln(1-X_C)}\right)\right)^{-1} \qquad (7.3)$$

τ varied between 0.69 and 1.55 s.

The solar power input through the aperture varied in the range 3.3-6.6 kW. The nominal reactor temperatures varied in the range 1296-1843 K. Chemical conversions for petcoke and steam after a single pass [defined by Equations (7.1) and (7.2)] reached up to 87% and 69%, respectively.

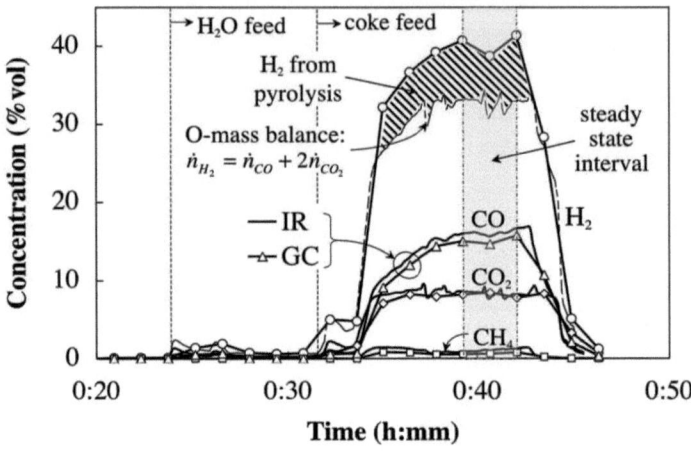

Figure 7.3: Off-gas composition of the reference experiment measured by GC and IR spectroscopy for the steam gasification of PD coke in the SynPet reactor.

7.3.1 Reference experiment

Gas composition, chemical conversion and temperatures for a representative solar run (number 20 of [102]) are shown in Figures 7.3, 7.4, and 7.5, respectively. In this run, the reactor was first heated to above 1400 K under an argon flow. Thereafter, reactants were introduced starting with steam and 8.5 minutes later coke, indicated by dashed vertical lines in Figures 7.3 and 7.4. Both reactants were fed during an interval of 12 min at a rate of 3.5 g petcoke/min and 9 g H_2O/min, corresponding to a H_2O/C molar ratio of 1.95.

Figure 7.3 shows the off-gas composition measured with the GC and IR detector. Approximate steady state is observed about 7 minutes after starting the coke feeder and maintained for another 5 minutes. Average values under approximate steady state condition are calculated for a time period of 3 minutes, indicated by gray boxes: $\dot{n}_{H_2} = 0.32$ mol/min, $\dot{n}_{CO} = 0.13$ mol/min, $\dot{n}_{CO_2} = 0.06$ mol/min, and $\dot{n}_{CH_4} = 0.0089$ mol/min. H_2, CO and CO_2 are primary products of the steam gasification reaction (Equation 3.2). Methane as well as the hatched part of the hydrogen signal are pyrolysis products of

7.3. RESULTS

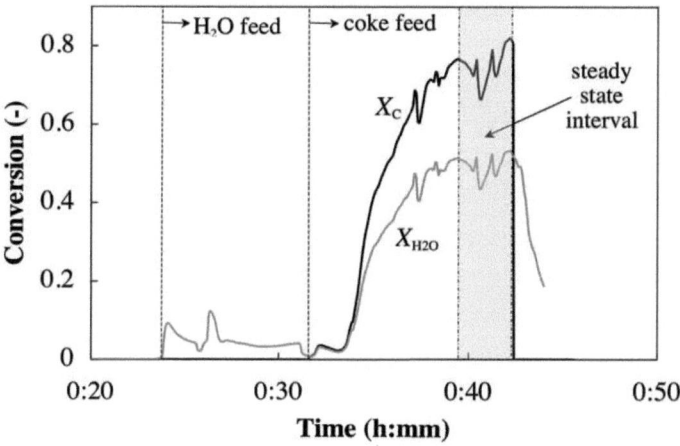

Figure 7.4: Carbon conversion, X_C, and steam conversion, X_{H_2O}, calculated with Equations (7.1) and (7.2), respectively, for the experiment in Figure 7.3.

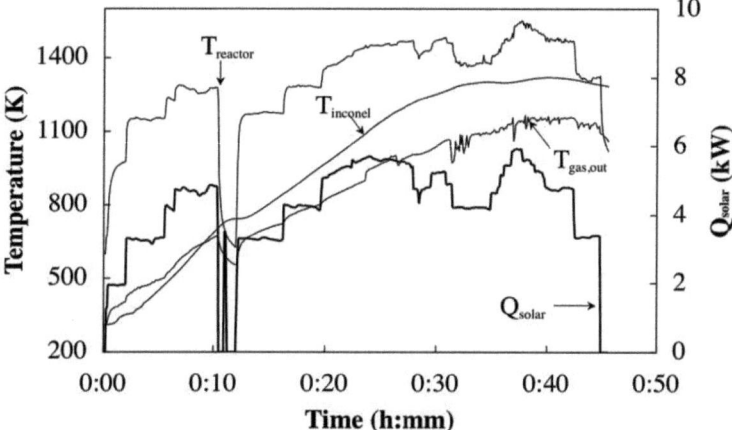

Figure 7.5: Pyrometer temperature, $T_{reactor}$, wall temperature, $T_{inconell}$, off-gas temperature, $T_{gas,out}$, and thermal input, Q_{solar}, of the experiment in Figure 7.3.

PD coke and disappear immediately as coke feeding is stopped. The hatched area in Figure 7.3 marks the difference between the hydrogen data reported by the GC and the hydrogen mass balance for the steam gasification reaction, $\dot{n}_{H_2} = \dot{n}_{CO} + 2 \cdot \dot{n}_{CO_2}$.

The chemical conversion of steam, X_{H_2O}, and carbon, X_C, is calculated with Equations (7.1) and (7.2). Figure 7.4 shows the conversion data of the reference experiment (cf. Figure 7.3). Since the reactor is operated with excess steam, carbon conversion is always higher than steam conversion. Measured average values under approximate steady state condition are: $X_{H_2O} = 51\%$ and $X_C = 78\%$

Figure 7.5 shows the reactor temperatures and the solar thermal input, Q_{solar}. $T_{reactor}$ is the cavity temperature measured with a solar-blind pyrometer. $T_{inconel}$ and $T_{gas,out}$ are the temperatures of the inconel shell and the off-gas stream after leaving the cavity, both measured with thermocouples. Averaged values of the present run during the steady state interval are $Q_{solar} = 4.6$ kW, $T_{reactor} = 1464$ K, $T_{Inconel} = 1318$ K.

7.3.2 SynPet experiments at PSI's solar furnace

Steam and carbon conversion

Figures 7.6, 7.7, and 7.8 show scatter plots of the steam conversion, X_{H_2O}, and the carbon conversion, X_C, which are considered as the primary performance indicators of the SynPet reactor. X_{H_2O} and X_C are reported as functions of experimental parameters being the molar carbon flow, $\dot{n}_{C,0}$, the steam to carbon stoichiometric ratio in the feed, $\dot{n}_{H_2O,0}/\dot{n}_{C,0}$, and the cavity temperature, $T_{reactor}$, measured with a solar-blind pyrometer. The experimental data was collected during 24 independent runs with the SynPet reactor at PSI's solar furnace. The reactor was operated at different temperatures with a feed consisting of PD coke ($d_{P,50} = 1.2$ μm) and steam at different molar ratios.

An analysis of the correlation between the chemical conversion of steam and carbon and the experimental parameters revealed that, as far as the 1400-1800 K temperature interval is considered, the reactor performance is only poorly affected by the nominal reactor temperature measured with the pyrometer. Figure 7.6 shows that above 1400 K the chemical conversion of the reactants is essentially independent of the pyrometer temperature. However, the chemical conversion is affected by the feed rates of coke and steam, apparent from Figure 7.7, and the H_2O/C stoichiometric ratio in the feed, apparent from Figure 7.8.

7.3. RESULTS

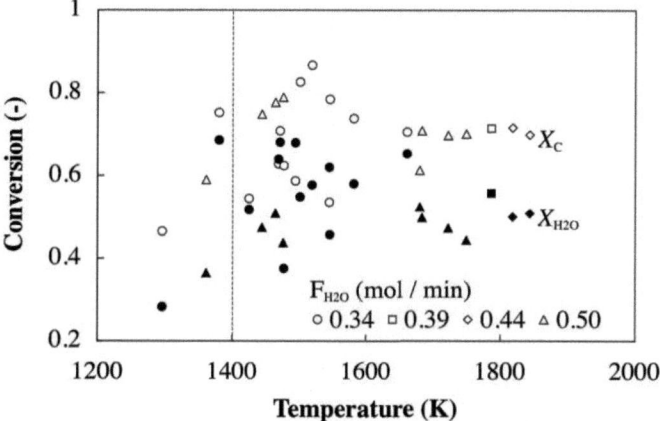

Figure 7.6: Steam and carbon conversion as function of the pyrometer temperature (cf. Figure 7.5) for the SynPet experiments with PD coke. $d_{P,50} = 1.2$ μm, $p_{H_2O} \in [0.54, 0.69]$ bar.

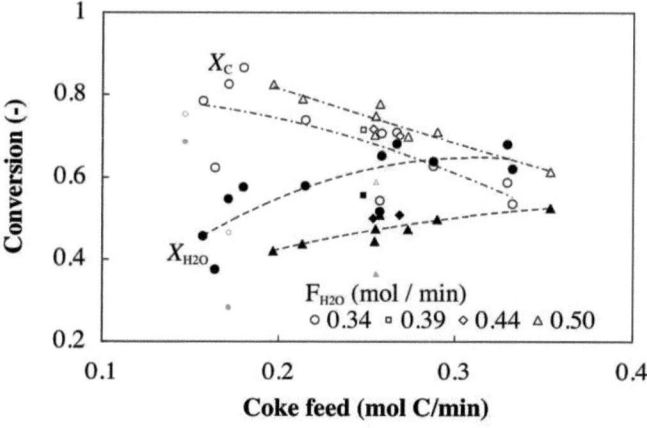

Figure 7.7: Steam and carbon conversion as function of the coke and steam feed rates for the SynPet experiments with PD coke. $d_{P,50} = 1.2$ μm, $p_{H_2O} \in [0.54, 0.69]$ bar, $T \in [1300, 1840]$ K.

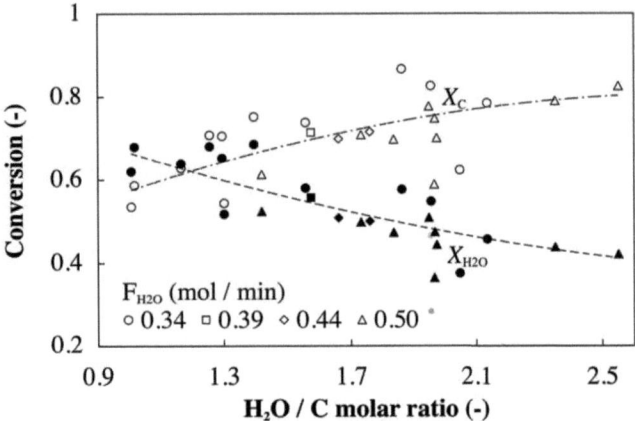

Figure 7.8: Steam and carbon conversion as function of the H_2O/C molar ratio (excess steam) in the feed for the SynPet experiments with PD coke. $d_{P,50} = 1.2\,\mu m$, $p_{H_2O} \in [0.54, 0.69]$ bar, $T \in [1300, 1840]$ K.

Because the temperature at the reaction site is known to have an important effect on the rate of the reaction and thereby on the chemical conversion, it is most probable that the relevant particle temperature of the reacting coke-steam mixture is not sufficiently correlated with the measured wall temperature. The reason is that coke particles passing at locations with high radiative flux densities, as for example the cavity center near the aperture, exhibit much higher particle temperatures than the wall temperatures recorded by the pyrometer. It is therefore the combination of the solar thermal input, Q_{solar}, and the characteristics of the radiative heat transfer in the particle cloud that controls the temperature at the reaction site rather than the wall temperature measured with the pyrometer.

Another reason for the poor correlation between the chemical conversion and the reactor shell temperature above 1400 K could be a possible mass transfer limitation at high temperatures. However, the particles used for the solar steam gasification experiments are very small ($d_{P,50} = 1.2\,\mu m$) making a mass transfer limitation in the fluid film surrounding the particles unlikely to happen. Chapter 8.4.2 gives further information about this issue.

Figure 7.7 shows the conversion of steam and carbon as a function of the molar feed rate of carbon, $\dot{n}_{C,0}$. The data are subdivided into 4 groups

7.3. RESULTS

according to the discrete water flow rates of 6 ('∘'), 7 ('□'), 8 ('◇'), and 9 ('△') g H_2O/min used in the experiments. The dashed lines (representing X_{H_2O}) and the dash-dotted lines (representing X_C) are polynomial trend lines based on data obtained from experiments with 6 and 9 g H_2O/min, respectively. Gray data points correspond to experiments with nominal reactor temperatures below 1400 K and were excluded from the calculation of the trend lines.

The conversion data for steam and carbon shown in Figure 7.7 are clearly correlated with both the carbon flow rate and the steam flow rate. Increasing the carbon feed at constant steam flow favors steam conversion and decreases carbon conversion while increasing the steam feed at constant carbon flow favors carbon conversion at the expense of steam conversion.

In terms of the H_2O/C molar ratio (see data presented in Figure 7.8), an increase from 1 (i.e. a stoichiometric feed) to above 2 (i.e. 100% excess steam) causes a raise of the carbon conversion from about 60% towards 80% and reduces the conversion of steam from 65% to 40%. The data for X_{H_2O} and X_C further follows a single trend indicated by dashed and dash dotted lines, respectively, making clear that the reactor performance depends rather on the H_2O/C stoichiometric ratio than on the absolute amount of carbon and steam in the feed.

Syngas quality

Besides the conversion of steam and carbon, the quality of the produced syngas characterized by the H_2/CO and CO_2/CO molar ratios is considered as a useful measure to describe the reactor performance. The Figures 7.9, 7.10, and 7.11 show the syngas quality obtained from the experiments with the SynPet reactor at PSI's solar furnace. The representation of the experimental data is analogous to the Figures 7.6, 7.7, and 7.8 discussed before.

The experimentally measured molar ratios of the product gas are in the order of $H_2/CO \in [2, 2.5]$ and $CO_2/CO \in [0.3, 0.5]$, which is notably higher than what is typically obtained from conventional autothermic gasification processes where heat is supplied by internal combustion (see [71], p. 64 et sq.).

In analogy to the results obtained for the chemical conversion of steam and carbon presented in Figure 7.6, the molar ratios of the product gases shown in Figure 7.9 are essentially constant at nominal reactor temperatures above 1400 K. At temperatures below 1400 K, both the H_2/CO molar ratio and the CO_2/CO molar ratio show an increase, indicating a higher amount of CO_2 as predicted by thermodynamics.

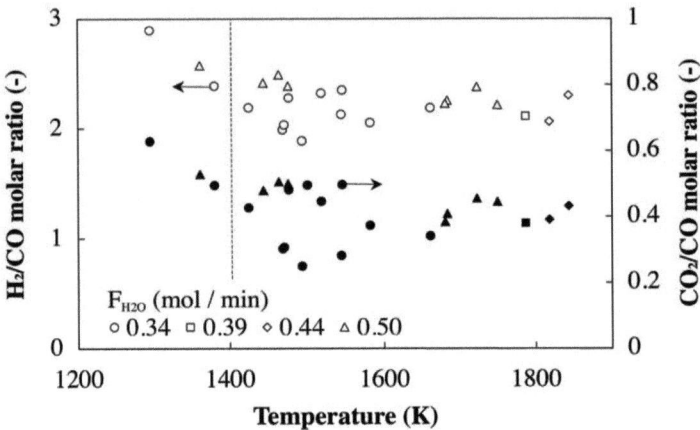

Figure 7.9: Product composition as function of the pyrometer temperature for the SynPet experiments with PD coke.
$d_{P,50} = 1.2\,\mu m$, $p_{H_2O} \in [0.54, 0.69]$ bar.

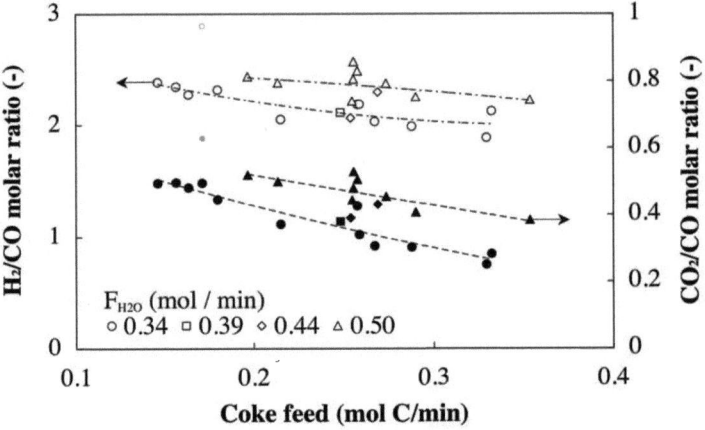

Figure 7.10: Product composition as function of the coke and steam feed rates for the SynPet experiments with PD coke.
$d_{P,50} = 1.2\,\mu m$, $p_{H_2O} \in [0.54, 0.69]$ bar, $T \in [1300, 1840]$ K.

7.3. RESULTS

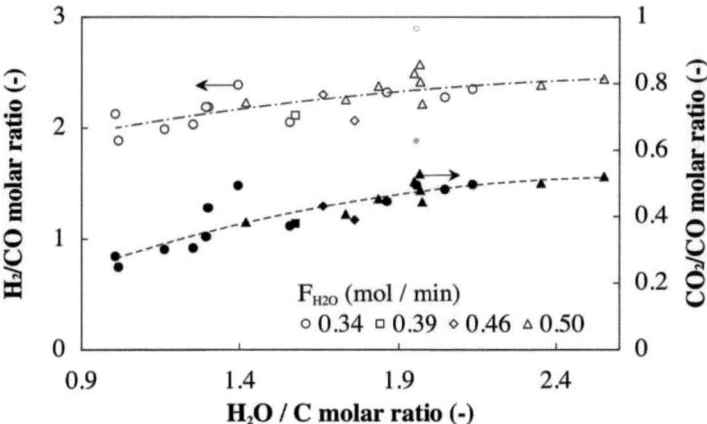

Figure 7.11: Product composition as function of the H_2O/C molar ratio (excess steam) in the feed for the SynPet experiments with PD coke. $d_{P,50} = 1.2\,\mu m$, $p_{H_2O} \in [0.54, 0.69]$ bar, $T \in [1300, 1840]$ K.

Both the absolute amount of steam and carbon fed to the reactor (cf. Figure 7.10) and the relative composition of the feed defined by the H_2O/C molar ratio (cf. Figure 7.11) affect the H_2/CO and CO_2/CO molar ratio of the product gas. Increasing the carbon feed rate at constant steam feed rate leads to a reduction of the characteristic molar ratios because less oxygen is available in the system. However, increasing the steam feed rate while maintaining the carbon feed rate constant favors the water gas shift reaction [Equation (3.6)], which produces additional hydrogen and carbon dioxide. The same is true for the correlation of the characteristic ratios of the syngas with the H_2O/C molar ratio in the feed.

Energy conversion efficiency

The energy conversion efficiency of the SynPet reactor was analyzed in detail by Z'Graggen et al. [102]: Two energy conversion efficiencies describing the reactor's thermal performance are defined:

- μ_1 is the portion of solar energy stored as chemical energy.

$$\mu_1 = \frac{Q_{chem}}{Q_{solar}} = \frac{X_C\,\dot{n}_C\,\Delta H_R|_{298K}}{Q_{solar}} \qquad (7.4)$$

- μ_2 is the portion of solar energy net absorbed, both as chemical energy and sensible heat (which potentially can be recovered). Further, μ_2 takes into account the heat required for steam generation.

$$\mu_2 = \frac{Q_{chem} + Q_{sensible}}{Q_{solar} + Q_{steam}} \tag{7.5}$$

The experimental efficiencies achieved during the SynPet experiments at PSI's solar furnace are $\mu_1 \in [5,9]\%$ and $\mu_2 \in [10,20]\%$. The complete energy balance for each experimental run in percent of the solar power input is found in [102]. Heat losses were found to be principally due to attenuation by the window and re-radiation through the aperture ($\approx 16\%$ of Q_{solar}) as well as conduction through the reactor walls ($\approx 67\%$ of Q_{solar}). The radiative losses showed to be strongly temperature dependent. A CPC is considered to minimize radiative losses by augmenting the input solar power flux and allowing the use of a smaller aperture for capturing the same amount of energy. Increasing the reactor temperature further results in a higher reaction rate and a higher amount of chemical conversion, which in turn results in higher energy conversion efficiencies.

7.4 Summary and conclusions

This chapter shows the design and testing of a solar chemical reactor for performing the solar steam gasification of petcoke to syngas. Solar experiments in a high-flux solar furnace were performed for a solar power input in the range 3.3-6.6 kW, a nominal reactor temperatures in the range 1296-1818 K, and a reactant mass flow rate in the range of 1.85-4.45 g petcoke/min and 3.68-9.04 g steam/min.

The chemical conversion reached up to 87% for petcoke and and up to 69% for steam. The syngas quality observed in the solar experiments was $H_2/CO \in [2, 2.5]$ and $CO_2/CO \in [0.3, 0.5]$, respectively, which is notably higher than what is typically obtained when heat is supplied by internal combustion of petcoke. The energy conversion efficiency, defined as the portion of solar energy net absorbed (both as chemical energy and sensible heat), was between 10 and 20%.

The chemical conversion of steam and carbon are considered as characteristic numbers for the reactor performance and their correlation with system parameters such as the nominal reactor temperature measured with a pyrometer, the H_2O/C molar ratio in the feed, and the steam and carbon feed flows is analyzed.

7.4. SUMMARY AND CONCLUSIONS

Increasing the molar feed rate of carbon, i.e. feeding more petcoke to the reactor at a constant steam flow, increases the conversion of steam and decreases both the conversion of carbon and the H_2/CO and CO_2/CO molar ratios, meaning that CO being favored on the expenses of CO_2. In opposition to increasing the carbon feed rate, increasing the steam feed rate has a positive effect on the carbon conversion and reduces the conversion of steam. The molar ratios of the product gases are both increased because the higher water partial pressure favors the water gas shift reaction [Equation (3.6)].

The correlation of chemical conversion and product gas composition with the nominal reactor temperature is poor and the measured shell temperatures can not be used to model the reactor performance. The reactor performance depends to a large extent on the severe conditions in the directly irradiated particle cloud characterized by a temperature maximum close to the reactor center and a 600 K temperature drop towards the colder reactor wall, as it will be presented in Chapter 8 elaborating on the reactor model. In the numerical model of the SynPet reactor, the temperatures of the reactants will be estimated by means of a heat transfer model based on a Monte-Carlo radiation heat transfer simulation.

Chapter 8

SynPet reactor modeling

8.1 Choosing the correct model

Depending on the reactor design, the flow behavior of a real chemical reactor is usually close to one of the two ideal flow patterns being *mixed flow* (MFR) and *plug flow* (PFR). The design of the SynPet reactor presented in Chapter 7 features a cylindrical cavity of 10 cm inner diameter and 20 cm length. Flow visualization experiments in a plexiglas model showed that gas and particles are entrained in a vortex flow with streamlines following a helical path. It is therefore reasonable to expect the SynPet reactor performance close to that of a PFR.

However, the vortex flow and turbulence caused by the nozzles for the injection of steam and purge gas at the reactor inlet are reasons for a deviation from the idealized plug flow pattern. It is therefore indicated to consider different reactor models that feature a higher complexity and enable the treatment of vessels with real flow behavior.

Several models are available that account for small deviations from the PFR pattern characterized by the spread of a tracer moving along the reactor [49, 78]:

- The *convection model for laminar flow* is used for reactors with low Reynolds numbers and laminar flow. It features a parabolic velocity profile and assumes that each element of fluid slides past its radial neighbor without interaction by molecular diffusion. The spread of the residence time distribution (RTD) which is characteristic for real flow is caused by radial velocity gradients only.

- The *axial dispersion model* considers a plug flow of fluid superimposed

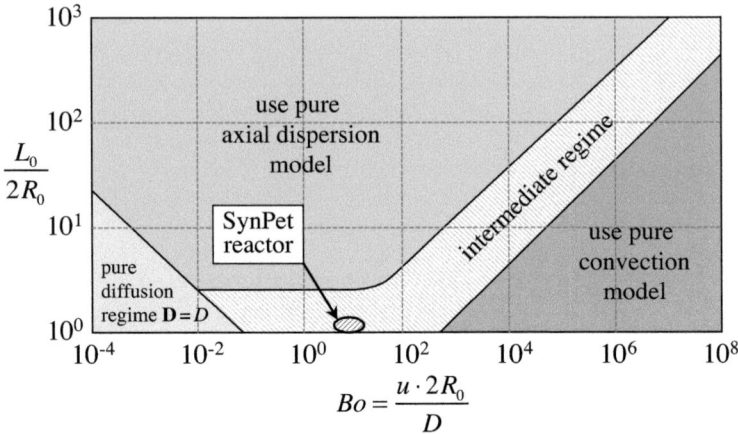

Figure 8.1: Map showing which flow model should be used [49]. The ellipse marks the location of the SynPet reactor with $Bo \approx 30$ and $L_0/2/R_0 = 2$.

by some degree of back-mixing caused by slippage and eddies. Back-mixing is modeled with longitudinal dispersion in the direction of the flow. The parameter that characterizes this diffusion-like process is the axial dispersion coefficient, $\mathbf{D}$ (m^2/s). The overall mixing behavior of the vessel is characterized by the dimensionless vessel dispersion number $\frac{\mathbf{D}}{uL}$.

Levenspiel [49] presents a standard procedure to decide which of the aforementioned models should be used. The procedure is based on a mathematical analysis of Ananthakrishnan et al. [3] reproduced in Figure 8.1. The chart allows to identify the flow regime by means of the dimensionless Bodenstein number

$$Bo = Re \cdot Sc = \frac{2R_0 u \rho}{\mu} \frac{\mu}{\rho D} = \frac{u 2 R_0}{D} \qquad (8.1)$$

and the vessel geometry, $L_0/2/R_0$.

The location of the SynPet reactor ($Bo \approx 30$ and $L_0/2/R_0 = 2$) is marked with a hatched ellipse in the intermediate regime between the axial dispersion and the convection model. For systems falling in the no-man's-land between regimes, Levenspiel [49] suggests to calculate the adjacent regimes and try averaging. Because gases are more likely to be in the dispersed regime than

8.2 One-dimensional axial dispersion model

The one-dimensional axial dispersion model assumes longitudinal dispersion in the direction of the flow (back-mixing) to account for the non-ideal characteristics of the flow. The primary parameter of the model is the dimensionless vessel dispersion number

$$\frac{\mathbf{D}}{uL} \tag{8.2}$$

which is a function of the axial dispersion coefficient, $\mathbf{D}$, the superficial velocity of the flow, u, and the length of the reactor vessel, L. The vessel dispersion number is a measure for the relative importance of convection and dispersion and characterizes the internal mixing which is represented by the spread of a tracer pulse traveling with the fluid through the reactor vessel. The limits to the real mixing behavior of the flow are given by the ideal flow patterns PFR ($\frac{\mathbf{D}}{uL} \to 0$) and MFR ($\frac{\mathbf{D}}{uL} \to \infty$), respectively.

Several authors pointed out the mathematical equivalence between one-dimensional molecular diffusion and axial dispersion [49, 16]. The mathematical treatment of a fluid undergoing fluctuations due to different flow velocities and due to molecular and turbulent diffusion is therefore identical to that of one-dimensional diffusion and convection. The only difference is that the diffusion coefficient D is replaced by the dispersion coefficient $\mathbf{D}$ having the same units but usually a different order of magnitude.

8.2.1 Governing equations

Conservation of mass. Using the aforementioned analogy between dispersion and molecular diffusion, the performance equation for a reactor with dispersed plug flow is obtained from the general equation of mass conservation for a multicomponent system incorporating chemical reaction [11]

$$\frac{\partial(\rho w_i)}{\partial t} = -\nabla m_i'' + r_i'' M_i \quad , \quad \left(\frac{\text{g}}{\text{m}^3 \cdot \text{s}}\right) \tag{8.3}$$

w_i is the dimensionless mass fraction and m_i'' is the mass flux of species i per unit reactor cross section consisting of a convection and a dispersion term:

$$m_i'' = \rho u w_i - \rho \mathbf{D} \nabla w_i \quad , \quad \left(\frac{\text{g}}{\text{m}^2 \cdot \text{s}}\right) \tag{8.4}$$

Assuming steady state conditions and substituting Equation (8.4) for m_i'' simplifies the mass conservation equation to

$$\nabla(\rho u w_i) - \nabla(\rho \mathbf{D} \nabla w_i) = r_i'' M_i \tag{8.5}$$

and for one-dimensional problems

$$\frac{d}{dx}(\rho u w_i) - \frac{d}{dx}\left(\rho \mathbf{D} \frac{dw_i}{dx}\right) = r_i'' M_i \tag{8.6}$$

Equation (8.6) is the one-dimensional mass conservation equation for species i in a multicomponent system. To obtain the concentration and velocity profiles along the reactor, $u(x)$ and $w(x)$ respectively, it is solved together with the continuity equation [66]

$$\frac{d}{dx}(\rho u) = 0 \tag{8.7}$$

$$\Rightarrow \rho u = const. \tag{8.8}$$

The reacting flow inside the reactor consists of a solid phase containing carbon (coke) and a fluid phase containing the gaseous reactants and products. To solve the differential mass conservation Equation (8.6) the solid phase is treated as pure carbon. The true composition and weight of the coke is only used to calculate the rate of reaction per unit coke weight.

Composition and density of the reacting flow. The overall composition of the two phase mixture is given in terms of the dimensionless mass fraction w_i

$$w_i = \frac{m_i}{\sum_j m_j} \tag{8.9}$$

The subscripts i and j denote the species in the mixture and are defined as follows: 1 = Ar, 2 = H_2O, 3 = H_2, 4 = CO_2, 5 = CO, 6 = CH_4 and 7 = C. Components 1 to 6 are present the gas phase and carbon represents the solid phase.

The composition of the gas phase required for the calculation of the reaction source term, $r_i'' M_i$, is given in terms of the molar fractions y_i for components 1 to 6

$$y_i = \frac{w_i M_i^{-1}}{\sum_j w_j M_j^{-1}} \tag{8.10}$$

The density of the two phase flow is defined in terms of the overall density and the local phase densities. The local phase densities for the gas and solid

8.2. ONE-DIMENSIONAL AXIAL DISPERSION MODEL

phase with respect to the single phase are

$$\rho_S = \rho_{coke} \sim 0.8 \cdot 10^3 \quad \left(\frac{\text{kg}}{\text{m}^3}\right)^1 \tag{8.11}$$

$$\rho_G = \frac{m_G}{V_G} = \frac{n_G \cdot \sum_i y_i M_i}{V_G} = \frac{P}{\mathcal{R}T} \sum_i y_i M_i \tag{8.12}$$

The overall density of the two phase flow with respect to the entire reactor volume is calculated using the local phase densities according to

$$\rho = \rho'_S + \rho'_G \tag{8.13}$$

$$\rho'_S = \frac{m_S}{V_{tot}} = \frac{m_S}{V_S + V_G} = \frac{m_S}{\frac{m_S}{\rho_S} + \frac{m_G}{\rho_G}} = \frac{w_S}{\frac{w_S}{\rho_S} + \frac{w_G}{\rho_G}} \tag{8.14}$$

$$\rho'_G = \frac{m_G}{V_{tot}} = \frac{m_G}{V_S + V_G} = \frac{m_G}{\frac{m_S}{\rho_S} + \frac{m_G}{\rho_G}} = \frac{w_G}{\frac{w_S}{\rho_S} + \frac{w_G}{\rho_G}} \tag{8.15}$$

8.2.2 Numerical methods

The one-dimensional axial dispersion model applied to the steam gasification of petcoke results in a system of 7 coupled ordinary differential equations plus one continuity equation. The nonlinear nature of the source term $r''_i M_i$ and the large number of equations prevents an analytical solution of the problem. The system of equations is therefore solved numerically using the finite volume method (FVM).

The concept of the FVM is to divide the calculation domain of a (partial) differential equation into a number of non-overlapping control volumes such that there is one control volume surrounding each grid point and solve the conservation equations in terms of the averaged values of the conserved quantities inside the control volumes. Figure 8.2 shows a schematic of a one-dimensional grid point cluster used to disctrize the axial reactor dimension (c.f. [66]).

The algebraic discretization equations are derived by integrating Equation (8.6) over the control volume:

$$(\rho u w_i)_e - (\rho u w_i)_w = \left(\rho \mathbf{D} \frac{dw_i}{dx}\right)_e - \left(\rho \mathbf{D} \frac{dw_i}{dx}\right)_w + r''_i M_i \Delta x \tag{8.16}$$

The subscripts e and w in Equation (8.16) denote the properties at the east and west face of the control volume.

---
[1]The bulk density of normal-quality petroleum coke is around 0.8 - 0.9 g/cm^3 [14].

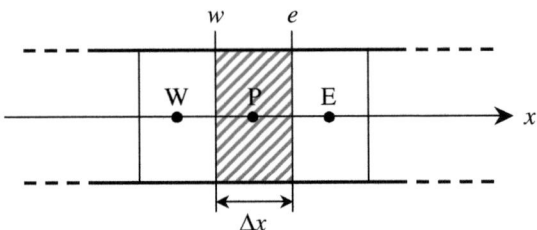

Figure 8.2: One-dimensional grid point cluster for the discretization. The hatched area marks the control volume of a generic point P and x is the axial reactor dimension.

Convection: Upwind scheme. The upwind scheme is used for the numerical computation of the convection expression. Since the flow in the reactor is downstream-only, i.e. $u(x) > 0 \; \forall \; x$, the concentrations at the faces of the control volume are approximated by the concentration of the upstream adjacent center point:

$$(w_i)_w = (w_i)_W \tag{8.17}$$
$$(w_i)_e = (w_i)_P \tag{8.18}$$

and the convective part of Equation (8.16) becomes

$$(\rho u w_i)_e - (\rho u w_i)_w = (\rho u w_i)_P - (\rho u w_i)_W \tag{8.19}$$

where P is the property at the control volume center and E and W represent the adjacent control volumes in the east and in the west.

Dispersion: Central difference scheme. The dispersion terms in Equation (8.16) are calculated assuming a piecewise linear profile to calculate the gradients and the flow properties at the control volume faces:

$$\left(\rho \mathbf{D} \frac{dw_i}{dx}\right)_w = \rho_w \mathbf{D}_w \left(\frac{dw_i}{dx}\right)_w = \rho_w \mathbf{D}_w \frac{(w_i)_P - (w_i)_W}{\Delta x} \tag{8.20}$$

$$\left(\rho \mathbf{D} \frac{dw_i}{dx}\right)_e = \rho_e \mathbf{D}_e \left(\frac{dw_i}{dx}\right)_e = \rho_e \mathbf{D}_e \frac{(w_i)_E - (w_i)_P}{\Delta x} \tag{8.21}$$

8.2. ONE-DIMENSIONAL AXIAL DISPERSION MODEL

with

$$\rho_w \mathbf{D}_w = \frac{1}{2}(\rho_W \mathbf{D}_W + \rho_P \mathbf{D}_P) \tag{8.22}$$

$$\rho_e \mathbf{D}_e = \frac{1}{2}(\rho_P \mathbf{D}_P + \rho_E \mathbf{D}_E) \tag{8.23}$$

leading to the following numerical scheme

$$\left(\rho \mathbf{D}\frac{dw_i}{dx}\right)_e - \left(\rho \mathbf{D}\frac{dw_i}{dx}\right)_w = \tag{8.24}$$
$$\frac{\rho_w \mathbf{D}_w}{\Delta x}(w_i)_W - \frac{\rho_e \mathbf{D}_e + \rho_w \mathbf{D}_w}{\Delta x}(w_i)_P + \frac{\rho_e \mathbf{D}_e}{\Delta x}(w_i)_E$$

Final discretized transport equation. Substituting the numerical schemes (8.19) and (8.25) into (8.16) yields the final discretized transport equation for the chemical species

$$(\rho u w_i)_P - (\rho u w_i)_W = \tag{8.25}$$
$$\frac{\rho_w \mathbf{D}_w}{\Delta x}(w_i)_W - \frac{\rho_e \mathbf{D}_e + \rho_w \mathbf{D}_w}{\Delta x}(w_i)_P + \frac{\rho_e \mathbf{D}_e}{\Delta x}(w_i)_E + r_i'' M_i \Delta x$$

Equation (8.26) can be expressed in a more general form as

$$a_P w_P = a_W w_W + a_E w_E + b \tag{8.26}$$

where the coefficients a_P, a_E, a_W and b take the form

$$a_W = F_W + D_w \tag{8.27}$$
$$a_P = F_P + D_w + D_e \tag{8.28}$$
$$a_E = D_e \tag{8.29}$$
$$b = r_i''(w_P) M_i \Delta x \tag{8.30}$$

with the location dependent variables

$$D \equiv \frac{\rho \mathbf{D}}{\Delta x} \tag{8.31}$$
$$F \equiv \rho u \tag{8.32}$$

Boundary conditions. For the control volumes 1 and N shown in Figure 8.3 at the beginning and the end of the reactor, Equation (8.26) takes a special form.

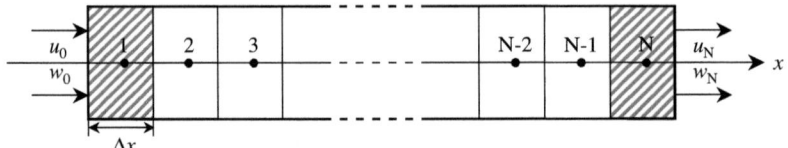

Figure 8.3: Discretization scheme of the reactor with the boundary elements 1 and N.

At the reactor inlet and outlet, the dispersion coefficients are zero in accordance with the closed vessel boundary conditions. As a consequence the dispersion flux at the reactor boundaries are zero as well and mass transfer from and to the outside appears by convection only. Additionally, the convective flux at the reactor inlet is known (u_0, w_0, ρ_0).

The coefficients of the discretization equation at the reactor inlet (element 1) take the form

$$a_0 = 0 \tag{8.33}$$
$$a_1 = F_1 + (D_1 + D_2)/2 \tag{8.34}$$
$$a_2 = (D_1 + D_2)/2 \tag{8.35}$$
$$b_1 = r_i''(w_1) M_i \Delta x + \underbrace{\rho_0 u_0 w_0}_{F_0} \tag{8.36}$$

and the respective coefficients at the reactor outlet (element N) are:

$$a_{N-1} = F_{N-1} + (D_{N-1} + D_N)/2 \tag{8.37}$$
$$a_N = F_N + (D_{N-1} + D_N)/2 \tag{8.38}$$
$$a_{N+1} = 0 \tag{8.39}$$
$$b_N = r_i''(w_N) M_i \Delta x \tag{8.40}$$

8.3 Axial dispersion number and residence time distribution

Two different methods are available to calculate the vessel dispersion number for a reactor with known flow properties: $\frac{D}{uL}$ can be obtained from correlations found in the literature (cf. [49]) or experimentally from the spread of

8.3. AXIAL DISPERSION NUMBER AND RTD

a tracer pulse moving through the reactor [49, 78]. The second method was used to calculate the dispersion number based on RTD measurements with a real size model of the SynPet reactor cavity.

Residence time distribution. The residence time distribution of a fluid flowing through a reactor vessel is characterized in terms of the distribution function (E-curve) or the cumulative distribution function (F-curve). The cumulative distribution function $F(t)$ denotes the share of the fluid that stayed inside the vessel for a time period shorter than t and is obtained experimentally by measuring the response of a step experiment with a nonreactive tracer.

$$F = \frac{c(t)}{c_0} \tag{8.41}$$

with
$$F(0) = 0 \tag{8.42}$$
$$F(\infty) = 1$$

The corresponding probability density function (E-curve) is the normalized time derivative of the F-curve

$$E = \frac{dF}{dt} \tag{8.43}$$

with
$$\int_0^\infty E \, dt = 1 \tag{8.44}$$

Both the E and the F curve can be represented in terms of the absolute time t in seconds or the dimensionless reduced time $\theta = \frac{t}{\tau}$.

The parameters used to characterize the continuous residence time distribution are the predictand μ and the variance σ^2. The predictand is defined as

$$\mu_t = \int_0^\infty t \cdot E(t) \, dt \tag{8.45}$$

with
$$\mu_t = \frac{V}{v} = \tau \tag{8.46}$$
$$\mu_\theta = 1 \tag{8.47}$$

where τ is the hydrodynamic residence time of the reactor in seconds. τ is the average time a fluid package stays in the reactor cavity.

The variance σ^2 is calculated according to

$$\sigma_t^2 = \int_0^\infty (t - \mu_t)^2 \, E(t) \, dt \qquad (8.48)$$

$$\text{with} \quad \sigma_\theta^2 = \frac{\sigma_t^2}{\mu_t^2} \qquad (8.49)$$

Axial dispersion number from the RTD. For a vessel with a known residence time distribution that meets certain preconditions (i.e. a roughly symmetrical, single peaked E-curve) the vessel dispersion number, $\frac{\mathbf{D}}{uL}$, can be calculated from the dimensionless variance, σ_θ^2. A detailed description of this concept is presented by Levenspiel [49] for the case of an infinite tube. Van der Laan [93] further improved the theory of Levenspiel by adding other geometries and boundary conditions such as vessels with closed boundary conditions (no dispersion across boundaries, $\mathbf{D}|_{x=0} = \mathbf{D}|_{x=L} = 0$):

$$\sigma_\theta^2 = \frac{2}{Pe^2} \left[Pe - 1 + e^{-Pe} \right] \qquad (8.50)$$

where Pe is the dimensionless Peclet number sometimes used equivalent to the inverse of the axial dispersion number,

$$Pe = \left(\frac{\mathbf{D}}{uL} \right)^{-1} \qquad (8.51)$$

Experimental RTD measurement.[2] Figure 8.4 shows the flow chart of the experimental setup used to measure the residence time distribution. A three dimensional plexiglass model having the same dimensions as the reactor cavity was used to analyze the gas flows inside the reactor. A pair of Bronkhorst HiTec flow controllers connected to a logical board were used to generate a concentration step in the feed line by switching the gas supply from Ar to CO_2. At the exit of the cavity, the CO_2 concentration is measured online by means of a Siemens Ultramat 22P infrared detector.

The experimental setup in Figure 8.4 consisted of three independent flow units, namely the upstream lines, the reactor cavity, and the downstream lines. Each unit has its own residence time distribution that affects the overall system response measured by the ULTRAMAT CO_2-analyzer. μ and σ^2 of the reactor cavity without upstream and downstream lines are obtained

---

[2] The experimental work was performed in the framework of a semester thesis by M. Meyer and M. Rusch [58].

8.3. AXIAL DISPERSION NUMBER AND RTD

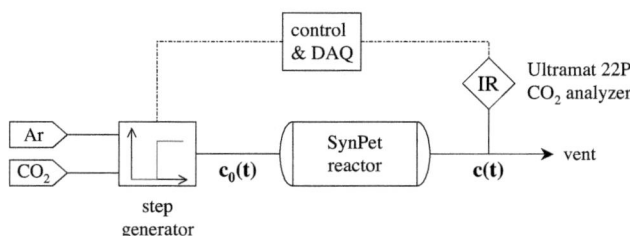

Figure 8.4: Flowchart of the experimental setup for the measurement of the RTD.

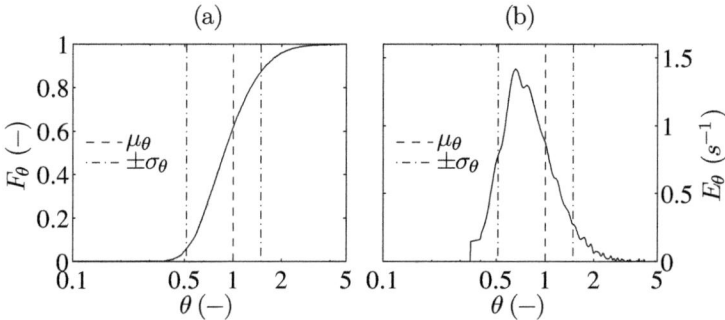

Figure 8.5: Experimentally determined residence time distribution of the SynPet reactor. (a) is the F-curve and (b) the corresponding E-curve.

Table 8.1: Distribution parameters and vessel dispersion number for the SynPet reactor.

experiment	F_{tot} (l_N/min)	μ_t (s)	σ_t^2 (s^2)	σ_θ^2 $(-)$	$\frac{D}{uL}$ $(-)$
F2027	6	8.63	11.33	0.152	0.0829
F2033	6	8.85	13.28	0.170	0.0936
F2041	8	6.57	5.94	0.138	0.0745

from the RTD data of the overall setup and using of the following expressions that relate μ and σ^2 for products of distribution functions [49, 80]:

$$\mu_{cavity} = \mu_{setup} - \mu_{pipes} \tag{8.52}$$

$$\left(\frac{\sigma_t^2}{\mu_t^2}\right)_{cavity} = \left(\frac{\sigma_t^2}{\mu_t^2}\right)_{setup} - \left(\frac{\sigma_t^2}{\mu_t^2}\right)_{pipes} \tag{8.53}$$

To determine the RTD of the SynPet reactor, a set of experiments was selected that visually showed a stable vortex flow during experiments with fog injection to visualize the stream lines. Figure 8.5 shows the experimentally determined F and E curves for the geometry of the SynPet reactor cavity calculated with Equations (8.41) and (8.43).

To calculate the distribution parameters μ and σ^2 for the reactor cavity, every run is repeated 3 times and the effect of the tubing is subtracted from that of the entire setup using Equations (8.52) and (8.53). The corresponding distribution parameters are listed in Table 8.1 for three different runs with a total flow rate of 6 and 8 l_N/min, respectively.

Equation (8.50) was used to calculate the axial dispersion number for the reactor cavity based on σ_θ. The results are listed in Table 8.1 together with the distribution parameters for every experiment. The obtained vessel dispersion numbers are in the range

$$0.075 < \frac{\mathbf{D}}{uL} < 0.094 \tag{8.54}$$

As a qualitative measure for the amount of back-mixing, Levenspiel [49] makes the following classification of flow regions with respect to $\frac{\mathbf{D}}{uL}$:

- $\frac{\mathbf{D}}{uL} < 0.01$: Small deviation from plug flow (PFR: $\frac{\mathbf{D}}{uL} \to 0$).

- $\frac{\mathbf{D}}{uL} > 0.01$: Large dispersion (MFR: $\frac{\mathbf{D}}{uL} \to \infty$).

The dispersion number measured for the SynPet reactor cavity is located in the second category comprising vessels with significant deviation from plug flow. It is therefore indicated to use a model that takes into account deviations from the PFR pattern caused by axial dispersion.

8.4 Chemical reaction

8.4.1 Reactivity correction for the solar experiments

First tests with the numerical model for the 5 kW SynPet reactor developed in Chapter 8.2 were done using the rate data obtained from the TG experiments in Chapter 5. However, the obtained results showed that the TG rate data can not reproduce the experimentally observed performance of the SynPet reactor. The carbon conversion predicted by the TG kinetics (Equation (4.83) with the rate constants from Table 5.7) was unexpectedly low.

In order to find an explanation for the performance of the process reactor during the solar experiments, additional TG runs were performed with coke residues collected in the off-gas cyclone (cf. Figure 7.2) after three different solar runs. The reactor samples were subsequently gasified in the TG applying the same conditions as previously used for the determination of the rate constants in Chapter 5.3.2 (i.e. a linear heating rate with $\beta = 20$ Kmin^{-1} from 470 to 1420 K and a reactive gas with 0.6 bar steam in argon).

The solid lines in Figure 8.6 represent the TG curves resulting from the gasification of PD coke residuals ($d_{P,50} = 1.2$ μm) that were collected after three solar experiments performed with $p_{H_2O} \in [0.6, 0.7]$ bar and $T \in [1580, 1680]$ K. The dashed lines are TG runs of the experimental campaign presented in Chapter 5: The raw sample (dashed line) is recognized by means of the pyrolysis step at around 900 K, the dash-dotted line is the TG curve of a sample that was pyrolyzed with 20 Kmin^{-1} and a maximum pyrolysis temperature of 1520 K in order to force thermal deactivation.

The data presented in Figure 8.6 clearly show that despite the exposition to temperatures far beyond the level of ≈ 1300 K which is identified in Chapter 5.2.2 as the threshold temperature towards significant thermal deactivation, the PD coke samples that passed through the hot SynPet reactor have not been deactivated. The carbon-steam reaction of the reactor residuals proceeds at similar temperatures as observed with raw PD coke samples but at higher rates. Further, the SynPet residuals do not show a pyrolysis step meaning that pyrolysis in the solar reactor is essentially complete.

The lack of thermal deactivation of the PD coke in the solar furnace is the result of the following circumstances:

- The residence time of the samples in the hot reactor cavity was $\tau \approx 1$ s. This is very short compared to the long duration of the TG experiments (≈ 1 h) used derive the rate data in Chapter 5. The characteristic time scale of the deactivation step is probably too big for the SynPet reactor.

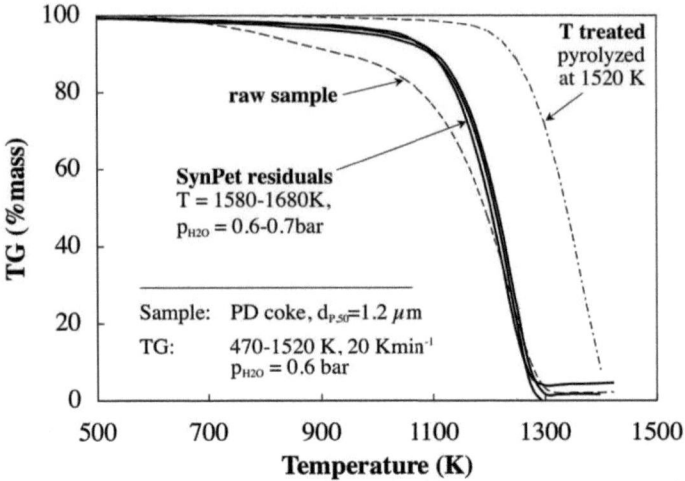

Figure 8.6: TG curves of PD coke residues collected after three runs in the solar furnace (solid lines) compared to the standard sample ('− − −' is a raw sample and '· − ·−' a sample subjected to pyrolysis in the TG at 1520 K).

- The heating rates in the SynPet reactor are very fast ($\beta_{SynPet} \approx 10^4$ to 10^6, [103]) and exceed those of the TG by several orders of magnitude.

- Gasification and pyrolysis in the SynPet reactor run simultaneously rather than in series as in the TG. Steam activation may therefore play a different role.

To account for the reactivity difference observed between raw PD coke samples and residual samples collected during the solar experiments, an Arrhenius type correlation constant, k_{react}, is introduced to adjust the rates presented in Chapter 5:

$$r_{C,SynPet} = k_{react} \cdot r_{C,TG} \quad (8.55)$$

with
$$k_{react} = a \exp\left(\frac{-b}{\mathcal{R}T}\right) \quad (8.56)$$

8.4. CHEMICAL REACTION

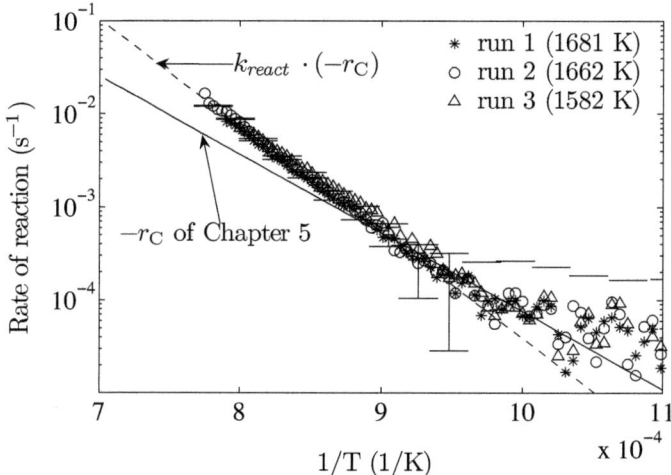

Figure 8.7: Rate of steam gasification, $-r_C$, with $p_{H_2O} = 0.6$ bar for three PD coke samples collected in the cyclone after the SynPet reactor (data points). The solid line represents the gasification kinetics from Chapter 5, the dashed line represents the kinetics from Chapter 5 multiplied with the reactivity constant, k_{react}.

Table 8.2: Numerical values for the parameters in Equation (8.56) obtained by minimization of the error between the left hand side and the right hand side of Equation (8.55).

constant	value
a (−)	$4.9796 \cdot 10^2$
b (J/mol)	$5.7144 \cdot 10^4$

Figure 8.7 is a plot of $\ln(-r_C)$ versus the inverse temperature showing the TG runs with SynPet residuals from Figure 8.6. The black solid line is the reaction rate calculated with Equation (4.83) and the Arrhenius parameters in Table 5.7 obtained for the gasification of PD coke with steam and carbon dioxide in Chapter 5. The deviation between data and model is apparent and not explicable with the error in the TG data (bars mark the experimentally determined error from Chapter 5.2.2). The dashed line is the result of Equation (8.56) using the parameters a and b of Table 8.2, obtained by minimization of the RMS absolute error between the left hand side and the right hand side of Equation (8.55). Error minimization is performed numerically with MATLAB's [37] Nelder-Mead Simplex algorithm generally referred to as unconstrained nonlinear optimization.

8.4.2 Mass transfer in the gas phase

In general, mass transfer limitation in the gas phase becomes important for large particles and high temperatures, respectively. To identify the particle size for which the influence of the gas phase diffusion resistance becomes important in the SynPet reactor, gas composition and carbon conversion along the reactor axis are calculated for various particle sizes using the model from Chapter 4.1.2. The results are then compared to the situation without gas phase diffusion limitation to see if there is an effect.

The reactor is assumed to operate isothermally at a cavity temperature of 1800 K. This temperature is chosen because it corresponds to the minimum temperature required to reach full conversion in a single pass with the feed characteristics of the experiments in Chapter 7. The results are shown in Figure 8.8 for particles with $R_P = 0.6$, 10, 75, and 150 μm. $R_P = 0.6$ μm is the particle size used for the experiments presented in Chapter 7 and $R_P = 75$ and 150 μm, respectively, are of particular interest for industrial applications.

Figure 8.8 shows the carbon conversion along the reactor axis (X_C, dashed line) together with the gas phase concentrations in the bulk ($y_{i,B}$, solid line) and at the particle surface ($y_{i,I}$, solid line with marker) as well as the concentration profile that would be obtained if diffusion in the gas phase was instantaneous (y_i, dotted line).

The 0.6 μm plot in Figure 8.8 represents the initial particle diameter used for the SynPet experiments in the PSI solar furnace. The respective data show that the gas phase concentrations at the particle surface and in the bulk are the same, perfectly matching the concentration profile for the case

8.4. CHEMICAL REACTION

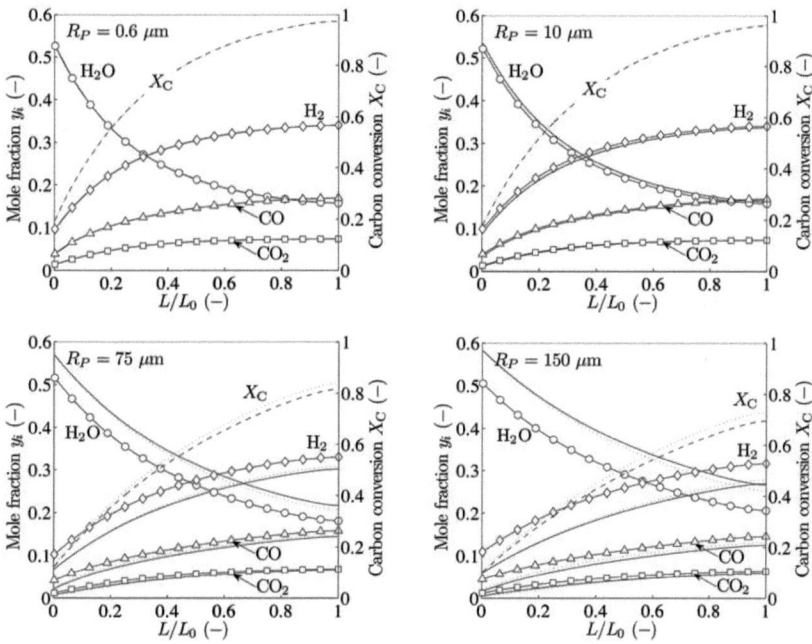

Figure 8.8: Gas composition (y_i) and carbon conversion (X_C) axial profiles for the SynPet reactor operated at 1800 K for four different particle sizes ($R_P = 0.6, 10, 75, 150\,\mu m$). The feed flow rates are those of the experiment presented in Figure 7.3.
Solid lines with marker represent the gas composition at the particle surface and solid lines without marker the corresponding bulk gas composition. Dotted lines represent the solution under the assumption of instantaneous diffusion in the gas film (no diffusion limitation).

of instantaneous gas phase diffusion. As the particles grow, the difference between bulk and interface concentration increases. The gradients of the products are negative in the direction away from the particle because of the reaction stoichiometry.

Due to the reduction of the steam concentration at the particle surface, the diffusion resistance in the gas phase lowers the carbon conversion by 3% for particles with $R_P = 150$ μm. However, the difference between the carbon conversion of a 0.6 μm and a 150 μm particle is 24% for the case of instantaneous gas phase diffusion. This points out that the performance loss observed for big particles is not a consequence of diffusion resistance in the gas phase, but rather the result of diffusion resistance inside the porous particle. Diffusion resistance in the particle is modeled with the particle effectiveness factor, η_P, which is ≈ 1 for 0.6 μm particles and drops to $\eta_P = 0.33$ for PD coke particles with $R_P = 150$ μm (cf. Figure 5.13).

Because diffusive mass transfer in the gas phase did not show any influence on the gasification performance of a 0.6 μm PD coke particle at all in the temperature interval required to attain full conversion, this effect is excluded from the numerical models of the SynPet reactor considered in the following chapters.

8.4.3 Reaction source term

The chemical model for the gasification of PD coke with steam considers two different reactions:

1. The pyrolysis of coke to form solid char and gaseous H_2 and CH_4.

2. The heterogeneous gas-solid reaction of the remaining char with H_2O and CO_2 to form H_2, CO and CO_2.

The source term in Equation (8.6), $r_i'' M_i$, is the rate of the chemical reaction with units gram per reactor volume and second. r_i'' represents the sum of the pyrolysis and the steam gasification reaction rates according to

$$r_i'' = \rho_S' \left(r_{i,P}' + a_0 \left(1 - X_C\right)^{2/3} r_{i,G} \right) \tag{8.57}$$

where $r_{i,P}'$ is the rate of the pyrolysis reaction with units mol per gram of solid (mol/g/s) and $r_{i,G}$ is the overall rate of the gasification reaction (mol/m²/s). The calculation of the pyrolysis and steam gasification rates, $r_{i,P}'$ and $r_{i,G}$, respectively, is elaborated in the following subsections.

8.4. CHEMICAL REACTION

Pyrolysis

The PD coke pyrolysis kinetics presented in Chapter 5.3.1 did not produce sufficiently fast reaction rates to predict the observed complete pyrolysis conversion in the SynPet process reactor (cf. Figure 8.6). The reason is the extremely different heating rate of the TG setup compared to that of the SynPet reactor. The heating rate is known to have a strong effect on the kinetics of the pyrolysis reaction [76], making the application of the TG kinetics to the SynPet reactor not feasible.

Therefore, pyrolysis is modeled as an instantaneous process taking place at the reactor entrance, $L = 0$. The pyrolysis products are solid char and gaseous hydrogen and methane at rates obtained from averaging the overall amount of pyrolysis gas of all SynPet experiments in the solar furnace. The overall amount of products released from the pyrolysis reaction, $\mathbf{r}'_i$, is

$$\int_0^\tau r'_{i,P}\, \delta(t)\, dt = \mathbf{r}'_i \tag{8.58}$$

$\delta(t)$ is the Dirac delta function $[\delta(0) = \infty$ and $\delta(t \neq 0) = 0]$ and the experimental $\mathbf{r}_i$ are:

$$\mathbf{r}'_{H_2} = 1.35 \cdot 10^{-2} \tag{8.59}$$
$$\mathbf{r}'_{CH_4} = 2.42 \cdot 10^{-3} \quad \text{(mol/g coke)} \tag{8.60}$$
$$-\mathbf{r}'_C = \mathbf{r}'_{CH_4} \tag{8.61}$$

The standard deviations of $\mathbf{r}'_i$ are $\sigma_{H_2} = 2.84 \cdot 10^{-3}$ and $\sigma_{CH_4} = 9.34 \cdot 10^{-4}$.

Steam gasification

The overall rate of the steam gasification reaction, $r_{i,G}$, is calculated from the intrinsic rate, $r_{i,intr}$, adjusted with the reactivity correction for the SynPet reactor, k_{react}, and the particle effectiveness factor, η_P. The overall rate is a function of temperature, gas phase composition and particle size.

$$r_{i,G} = \eta_P \cdot k_{react} \cdot r_{i,intr} \tag{8.62}$$

$r_{i,intr}$ is the intrinsic gasification rate for species i calculated with the Langmuir-Hinshelwood-type rate laws (4.77)-(4.80) derived from the extended mechanism in Chapter 4. Rate constants are calculated using the rate data obtained from gasification experiments with H_2O-CO_2 mixtures in the TG setup presented in Chapter 5. The corresponding Arrhenius parameters are found in Table 5.7.

To account for the reactivity difference observed for PD coke gasified with the TG and samples gasified in the solar reactor, the intrinsic rate of reaction is corrected with the temperature dependent proportionality factor, k_{react}, defined by Equations (8.55) and (8.56).

The particle effectiveness factor, η_P, is calculated with the experimental correlation defined by Equation (5.4) and using the parameters in Table 5.4. The very small coke particles used for the SynPet experiments in the PSI solar simulator have a effectivity close to 1: $\eta_P(R_P = 0.6 \mu m) = 0.99$.

Since the TG setup is considered as a differential reactor with negligible concentration gradients in the gas phase, the respective experiments could not provide data for the rate constant K_3 associated with CO_2-production by the water gas shift reaction (3.6). This problem is circumvented by the use of the activation energy for K_3 obtained from the gasification experiments in the quartz fluidized bed reactor and a pre-exponential factor that was determined empirically by minimization of the sum of the relative errors between the measured and simulated off-gas compositions y_{H_2}, y_{CO}, and y_{CO_2}.

Thus,

$$K_{3,TG} = 1.149 \cdot 10^{-7} \cdot \exp\left\{\frac{176\,100}{\mathcal{R}\,T}\right\} \tag{8.63}$$

8.5 Radial temperature profiles

Temperatures in the SynPet reactor were experimentally determined at various locations using different measurement techniques: Type K thermocouples connected to the reactor shell were used to measure the wall temperature. Type S thermocouples in the off-gas stream and inside a closed ceramic tube inserted to the cavity measured off-gas and cavity temperature. Temperature measurement by thermocouples was complemented with a solar blind pyrometer measuring the temperature inside the reactor cavity from the radiation leaving the reactor through the front window [102].

Both the SynPet reactor model of the present thesis and the more detailed model of Z'Graggen [102] unveiled that the reactor temperature experimentally determined by thermocouples at various locations as well as by pyrometry can not be used to explain the experimentally observed reactor performance. Neither the order of magnitude nor fluctuations of the chemical conversion could be adequately predicted with models using experimental wall temperature as the temperature of the reacting carbon-steam system.

To overcome the lack of suitable temperature data for the gas-particle mixture exposed to highly concentrated solar radiation, Z'Graggen developed

8.5. RADIAL TEMPERATURE PROFILES

a detailed model for the heat transfer inside the SynPet reactor featuring a 3D path length Monte-Carlo ray tracer [100]. The model was used to calculate the temperature distribution of a static particle cloud in the reactor cavity based on the experimental conditions of the SynPet experiments (i.e. chemical composition w_i and Q_{solar} having the characteristics of the PSI solar furnace) presented in Chapter 7, taking into account radiation, convection and conduction effects as well as chemical reaction.

Similar calculations with flowing systems [102] showed that gas and particle temperatures inside the SynPet reactor reach a maximum close to the coke inlet and stay essentially constant as the particle cloud passes the cavity towards the exit. It is therefore reasonable to assume that the reactor is isothermal in the axial direction and has a radial temperature profile.

Figure 8.9 shows the radial distribution of the particle temperature at the reactor inlet for the conditions of the SynPet experiments in the solar furnace, presented in Chapter 7. The data are calculated with the Monte-Carlo radiation model [100]. The calculated profiles are bell shaped with a pronounced maximum at the cavity center. Temperature gradients in the radial direction are in the order of 600 K per 10 cm, whereas axial gradients are negligible, pointing out that the reactor performance is controlled by the radial temperature distribution.

Figure 8.10 shows the peak temperature ('△'), the integral mean temperature ('○'), and the wall temperature ('▽') of the coke particles extracted from Figure 8.9. The experimental runs are ordered by increasing pyrometer temperature ('∗'). The integral mean temperature over the reactor cross section is calculated according to

$$\bar{x} = \frac{2}{R^2} \int_0^{R_0} x(R) \cdot R \, dR \qquad (8.64)$$

Despite a monotonic increase of the experimentally determined pyrometer temperature ('∗'), the calculated mean particle temperature ('○') is essentially constant in all of the runs.

The exponential temperature dependence of the reaction rate originating from Equation (4.36) prevented the use of an isothermal reactor model as it was considered in Chapter 8.2. The use of a radial mean temperature for the performance calculation underestimates the substantial contribution of the hot reactor center where the Monte-Carlo model predicts temperatures as high as 2300 K. Considering the particle temperatures predicted by the Monte-Carlo model, the innermost parts of the reactor easily reach 100% conversion, whereas the cold regions close to the wall yield very low conversions (cf. Chapter 8.6).

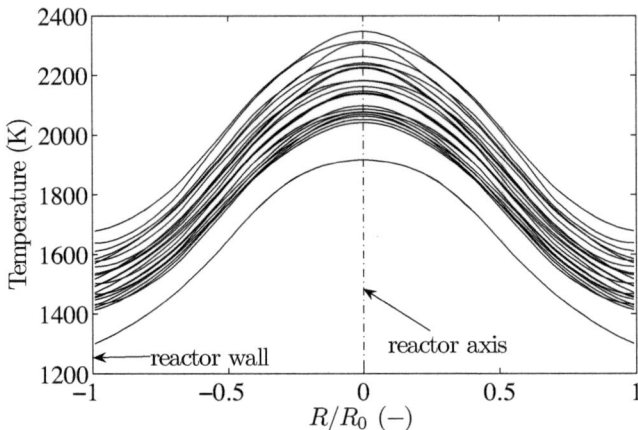

Figure 8.9: Radial distribution of the particle temperature for the conditions of the SynPet experiments in the solar furnace (cf. Chapter 7). The data are calculated with the Monte-Carlo radiation model [100].

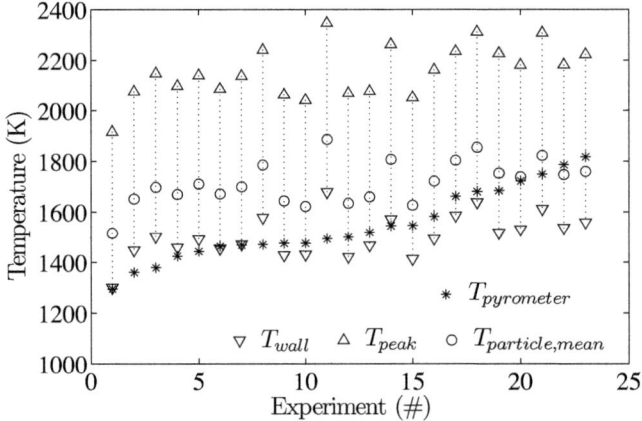

Figure 8.10: Peak temperature ('$\triangle$'), integral mean temperature ('$\circ$'), and wall temperature ('$\triangledown$') of the coke particles extracted from Figure 8.9. The experimental runs are ordered by increasing pyrometer temperature ('$*$').

8.6. RESULTS AND EXPERIMENTAL VALIDATION

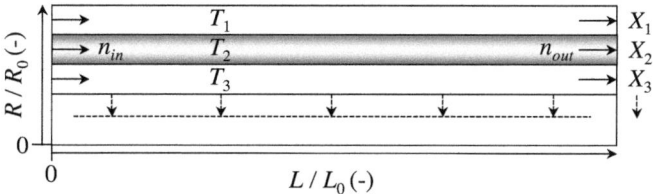

Figure 8.11: Schematic showing the radial discretization of the SynPet reactor with respect to temperature.

To address this issue, the reactor volume is subdivided into radial segments as shown in the schematic of Figure 8.11. The radial segments are considered non-interacting as it is observed in the case of segregated flow, being a reasonable assumption for tubular flow reactors with stream line flow [78]. Depending on its radial position, every segment is allocated with a characteristic temperature value obtained from the respective profiles in Figure 8.9. To obtain the overall performance at the reactor outlet under the assumption of a specific radial temperature profile and feed flow rate, the one-dimensional reactor model of Chapter 8.2 is solved for every segment to obtain the radial composition and flow rate distribution of the reacting mixture at the reactor outlet. The overall properties are then obtained by taking the integral mean [Equation (8.64)] over the reactor cross section.

8.6 Results and experimental validation

8.6.1 Reference experimental run

Figures 8.12 to 8.15 show the model results obtained for a single experiment. The presented run is the reference experiment of Chapter 7.3 presented in Figures 7.3 to 7.5, where reactants are fed at 3.5 g petcoke/min and 9 g H_2O/min, corresponding to a molar H_2O/C ratio of 1.95. The measured overall carbon conversion was $X_C = 78\%$.

Because of the big temperature differences exceeding 600K in the radial direction and the exponential temperature dependence of the reaction rate, all flow properties calculated at the reactor outlet ($L = L_0$) are strongly temperature dependent.

Figure 8.12 shows the radial temperature profile of the coke particles, the

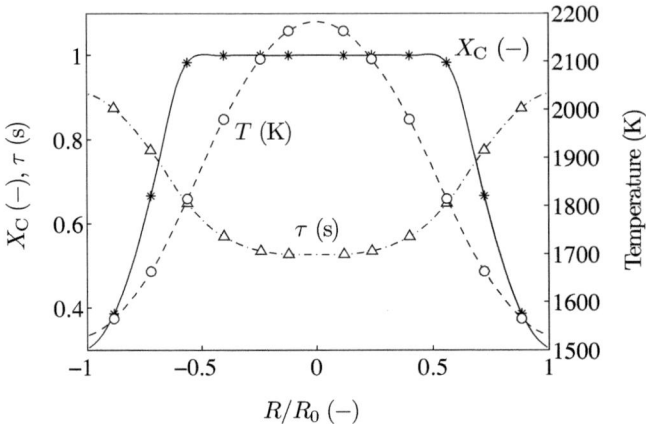

Figure 8.12: Radial variation of carbon conversion, residence time and temperature at the reactor outlet, calculated for the reference experiment in Chapter 7 (cf. Figures 7.3-7.5).

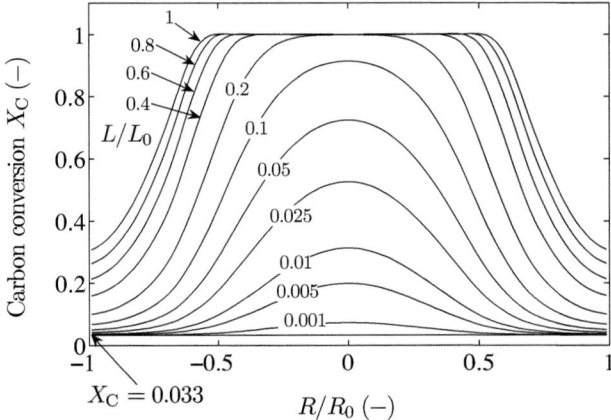

Figure 8.13: Radial variation of the carbon conversion at different positions along the reactor axis for the reference experiment. The parameter is the axial position L/L_0.

8.6. RESULTS AND EXPERIMENTAL VALIDATION

carbon conversion and the residence time as function of the dimensionless radial position, R/R_0, at the reactor outlet. The temperature maximum of 2192 K is located at the cavity center. Carbon conversion reaches completion at temperatures above $\approx$ 1800 K. Since higher particle temperatures are achieved on the inner half of the reactor radius, this region yields 100% conversion of the coke, whereas it drops significantly in the vicinity of the reactor wall reaching a minimum in the order of $X_C \approx 0.3$. As the cross sectional area is proportional to the square of the radius, the contribution of the peripheral area is more important than that of the central area where full conversion is achieved.

The radial distribution of the residence time, τ, is strongly affected by conversion and thermal expansion. Due to the stoichiometry of the gasification reaction, the amount of gas is almost doubled per mole of carbon reacting, leading to higher flow rates. The flow rates are further increased due to the thermal expansion being higher in the hot zone at the cavity center. The residence time τ is minimal at the cavity center where gas production and thermal expansion have a maximum.

Figure 8.13 shows the radial distribution of the carbon conversion and its change with the axial reactor position between the reactor inlet and outlet. The parameter is the axial reactor coordinate, denoted by the dimensionless reactor length L/L_0. $L/L_0 = 0$ is the reactor inlet and $L/L_0 = 1$ the reactor outlet. The carbon conversion at the reactor inlet is 3.3% for any radial position, because of the pyrolysis reaction assumed to take place in the first control volume, converting a small amount of the coke to methane and hydrogen. Moving the axial position in the direction of the outlet results in higher carbon conversions, especially near the hot cavity center where 50% of the coke is gasified at $0.025L_0$ and 100% between 0.1 and $0.2L_0$. However, the carbon conversion in the reactor segments close to the reactor wall increases only little between reactor inlet and outlet, despite the higher residence time close to the wall compared to the reactor center.

Figures 8.14 and 8.15 show the modeled gas composition, y_i, and carbon conversion, X_C, inside the cylindrical reactor cavity at the reactor outlet and at the 1800 K isotherm with $R/R_0 \approx 0.6$, respectively.

The radial concentration profile at the reactor outlet is shown in Figure 8.14. Composition of the product gas depends on chemical conversion and temperature. Close to the reactor wall where conversion is low, the weight fraction of the reactants H_2O and C has a maximum whereas steam gasification products H_2, CO and CO_2 are mainly found in the hot reactor center. Besides the conversion maximum at the reactor center, the high tempera-

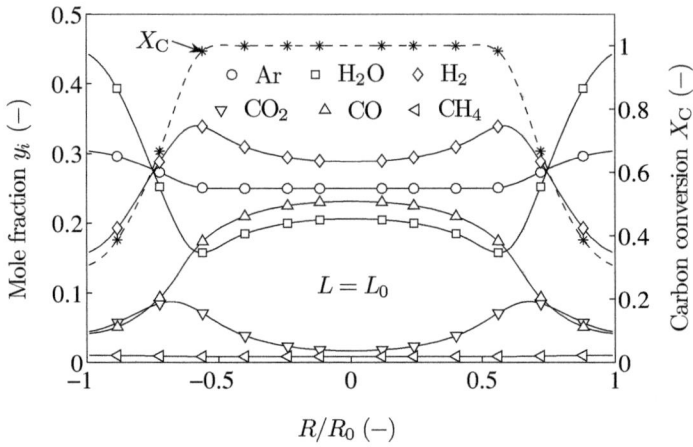

Figure 8.14: Radial profiles of the gas composition, y_i, and the carbon conversion, X_C, of the reference experiment at the reactor outlet ($L = L_0$).

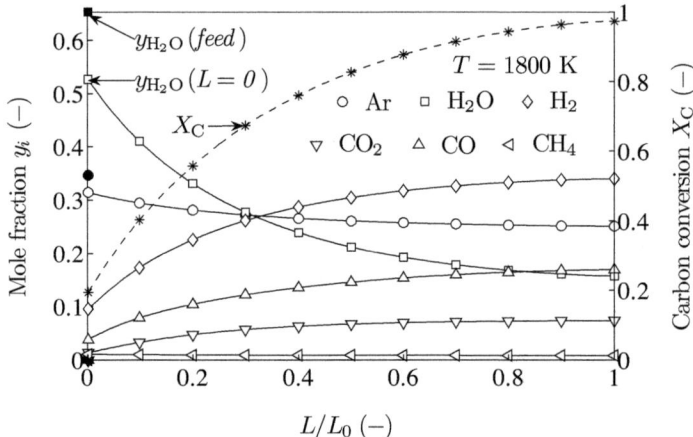

Figure 8.15: Axial profiles of the gas composition, y_i, and the carbon conversion, X_C, of the reference experiment at $R/R_0 \approx 0.6$ having a calculated temperature of 1800 K.

8.6. RESULTS AND EXPERIMENTAL VALIDATION

tures have an effect on the CO/CO_2 ratio of the syngas: A CO_2-rich syngas is produced close to the cold wall while the hot center produces mainly CO and H_2.

Figure 8.15 is a plot of the axial concentration profile at a distance $R/R_0 \approx 0.6$ from the cavity center, corresponding to a nominal reactor temperature of 1800 K. The feed composition ($w_{Ar} = 0.464$, $w_{H_2O} = 0.393$, $w_C = 0.143$) is indicated in terms of the molar fraction by black symbols (e.g. '■' for steam) on the y-axis, representing the cavity front end at $L = 0$. The respective gas composition along the cavity is marked with outlined symbols ('□').

Due to the dispersion term in the mass conservation Equation (8.6), the composition of the reactants inside the cavity at $L = 0$ differs from the respective feed composition. The axial dispersion causes a smoothening of the axial concentration gradients and an offset between the concentration in the feed and the concentration inside the reactor. The size of the offset depends on the axial dispersion number and becomes maximal for mixed flow behavior. The offset is further negative for reactants and positive for products and continuously reduces the reactor performance as the flow pattern changes from plug flow to mixed flow.

Between the reactor inlet and the reactor outlet, the composition of the reactants and products changes due to the chemical reactions defined by to Equations (4.77) to (4.80). The observed concentration profile is similar to what is expected for an ideal plug flow reactor. At $L = L_0$, essentially all carbon is converted to syngas resulting in a gas mixture consisting of syngas, argon, excess steam, methane from the pyrolysis reaction, and traces of unreacted carbon (petcoke).

8.6.2 SynPet experiments in the PSI solar furnace

This section presents an experimental validation of the SynPet reactor model elaborated in the previous sections. To estimate the accuracy of the model, the calculated product gas compositions and chemical conversions of carbon and steam are compared with experimental data from the experiments performed at PSI's solar furnace (cf. Chapter 7).

Product gas composition

Experimentally measured and modeled concentration data for the SynPet experiments in the solar furnace are presented in Figure 8.16. The plots show the molar composition of the relevant off-gas composition sorted according

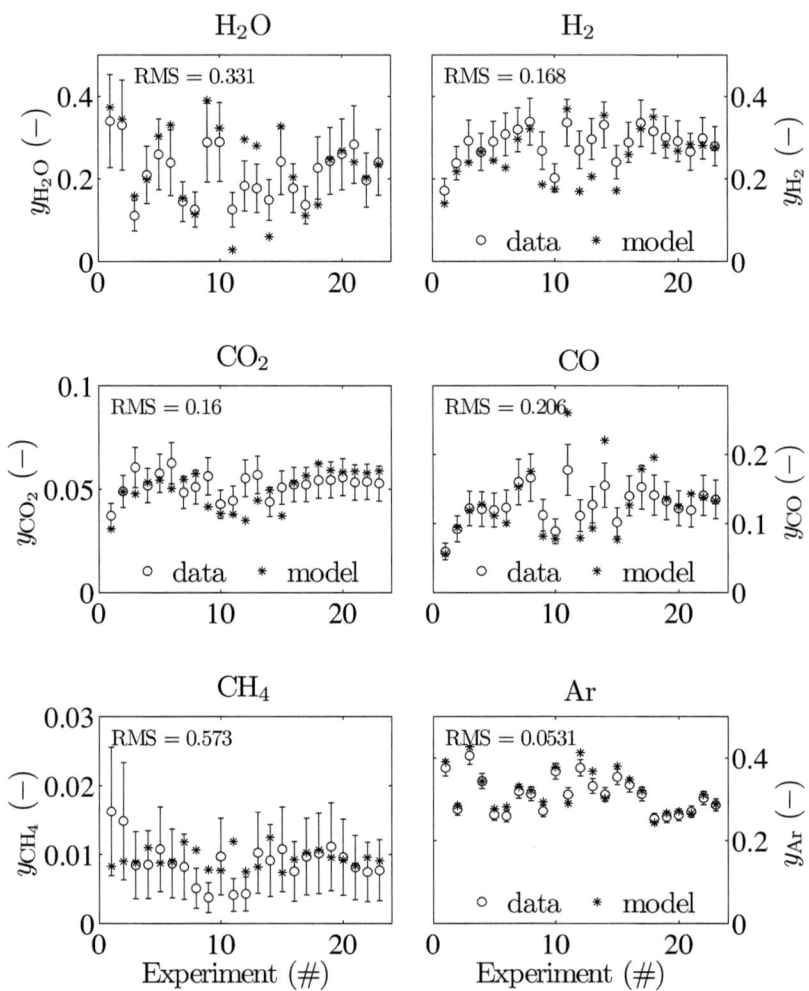

Figure 8.16: Experimental and simulated off-gas composition, y_i, for the SynPet experiments in the solar furnace (cf. Chapter 7).

8.6. RESULTS AND EXPERIMENTAL VALIDATION

to the nominal reactor temperature. Experimental data is complemented with error bars marking the root mean square (RMS) relative error between experiment and model.

The SynPet reactor model is capable to reproduce the concentration of the primary steam gasification products, i.e. H_2, CO and CO_2, with an average relative error of 17.8%. Less accuracy is obtained for H_2O and CH_4 with a relative error of 33.1 and 57.3%. The larger error of H_2O is explained with the fact that the steam concentration in the off-gas was not measurable experimentally and had to be calculated via oxygen mass balance.

$$\dot{n}_{H_2O} = \dot{n}_{H_2O,0} - \dot{n}_{CO} - 2\dot{n}_{CO_2} \qquad (8.65)$$

CH_4 concentration on the other hand is the product of a simplified treatment of the pyrolysis reaction and was calculated by taking the average methane production per unit coke weight and assuming that the pyrolysis reaction takes place in the first control volume. It is thus the result of a rather coarse calculation and further neglects possible de-carbonization reactions occurring at high temperatures.

The highest accuracy is found with an error of only 5.3% for Ar, which is the inert component and does not participate in the reaction.

Chemical conversion

Figures 8.17 and 8.18 show the chemical conversion of carbon, X_C, and steam, X_{H_2O}, respectively. Experimental data ('*') from the SynPet campaign in the solar furnace (cf. Chapter 7) and data from the numerical model ('o') in the order of increasing nominal reactor temperature, T_{pyro}, measured with the pyrometer (dashed line). The RMS of the relative error between experimental and modeled data is 20.8% for X_C and 22.0% for X_{H_2O}.

Comparing the trend in the conversion data with that of the nominal reactor temperature, T_{pyro}, it is obvious that neither of the conversions are directly depending on the nominal reactor temperature. The latter is steadily increasing over the experimental data series, whereas conversion stayed rather constant. In fact, both measured and modeled conversion data follow more closely the trend in the integral mean temperature based on temperature data calculated with the static Monte-Carlo model and previously shown in Figure 8.10.

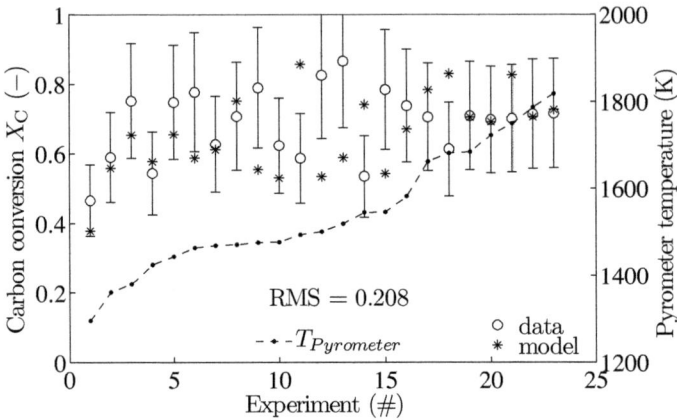

Figure 8.17: Experimental and simulated carbon conversion, X_C, for the SynPet experiments in the solar furnace (cf. Chapter 7).

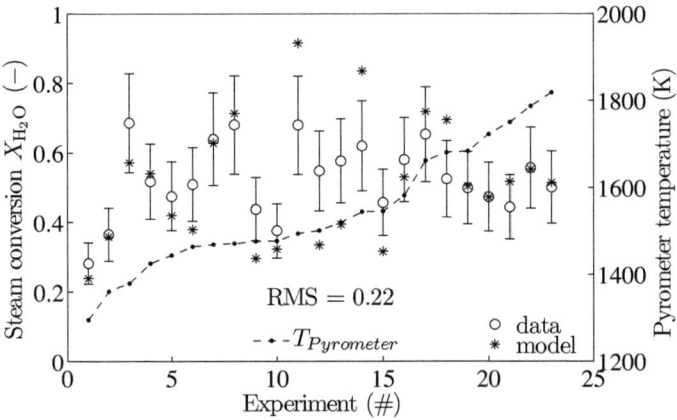

Figure 8.18: Experimental and simulated steam conversion, X_{H_2O}, for the SynPet experiments in the solar furnace (cf. Chapter 7).

8.7 Summary and conclusions

This chapter treats the numerical modeling of the SynPet 5 kW process reactor. A one-dimensional mathematical model is developed based on the *axial dispersion model*. The model accounts for non-ideal flow behavior by superimposing plug flow with axial velocity fluctuations due to different flow velocities and due to molecular and turbulent diffusion. The intensity of back-mixing is characterized by the dimensionless vessel dispersion number $\frac{D}{uL}$ being a measure for the spread of a tracer pulse passing through the vessel. The vessel dispersion number is determined experimentally doing residence time measurements with a reactor model having the same size and similar flow properties.

The one-dimensional dispersion model produces a system of nonlinear coupled ordinary differential equations that is solved with the *finite volume method* using upwind and central difference schemes for the discretization of the convection and dispersion term, respectively. Since the dispersion term prevents a direct solution, the system of equation is solved iteratively for a given temperature and feed composition.

Tests with an isothermal implementation of the model unveiled that the experimental cavity temperatures measured with thermocouples and pyrometer can not be used to explain the observed reactor performance. The model is therefore extended by radial temperature profiles calculated with a Monte-Carlo radiation model [102], taking into account radiation, convection and conduction effects as well as the heat of reaction. The static Monte-Carlo model predicts significant radial temperature gradients in the order of 600 K and peak temperatures exceeding 2200 K inside the reactive particle cloud providing reasonable results with respect to the overall reactor performance.

The intrinsic rate of the chemical reaction is calculated using the Langmuir-Hinshelwood rate laws (4.77)-(4.80) derived for the oxygen exchange mechanism in Chapter 4 with the rate data obtained for the H_2O-CO_2 gasification in the thermobalance presented in Chapter 5. Additional gasification experiments in the thermobalance performed with coke residuals from the SynPet campaign showed a systematic deviation of the intrinsic gasification rates of raw coke samples and samples that passed the solar reactor. The difference is explained with the much lower residence time of the coke in the process reactor compared to the TG and is corrected with an Arrhenius type correlation parameter.

The overall rate of the chemical reaction is calculated according the models presented in Chapter 4.1.1 and 4.1.2, taking into account mass transfer in the solid particle and in the surrounding gas phase, respectively. Mass transfer limitation inside the particle is considered in the form of an experimental correlation relating the particle effectiveness η_P with the particle diameter. Mass transfer of products and reactants between the particle surface and the bulk phase is modeled as steady diffusion from a sphere into a stagnant gas, assuming particle entrainment with negligible gas particle relative velocity. Gas phase mass transfer was found to have no effect in the case of a particle with $R_P = 0.6$ μm as it was used for the SynPet campaign, up to temperatures as high as 1800 K and carbon conversions up to 100%. As a consequence, only mass transfer limitations inside the particle are included in the model and the net gas production is calculated from the sum of the pyrolysis and reactive gasification according to Equations (8.57) and (8.62).

Modeling results reported for the reference experiment at the reactor outlet show a strong radial variation of overall composition, chemical conversion and flow rates. Close to the reactor center, where the temperature profiles show pronounced maxima exceeding 2200 K, 100% carbon conversion is achieved. The product gas has a low a low CO_2/CO ratio because of the high temperatures and residence time is minimal due to gas production and the thermal expansion. In the peripheral reactor regions the static Monte-Carlo model predicts lower temperatures between 1400 and 1700 K yielding only partial carbon conversion and a higher CO_2 content. Axial concentration profiles are similar to what is expected for a plug flow reactor with the exception of a concentration offset at the reactor inlet caused by dispersion.

The overall accuracy of the numerical model for the SynPet 5 kW process reactor is rated in terms of the RMS relative error between experimentally determined off-gas composition and carbon and steam conversion. The SynPet reactor model is capable to reproduce the concentration of the primary steam gasification products being H_2, CO and CO_2 with an average relative error of 17.8%. The RMS relative error of steam and carbon conversion are 20.8 and 22%. There are several reasons for the aforementioned notable relative errors:

- The model is considering chemical reaction and mass transfer only. The energy balance in particular is not included in the model as there is an important contribution of radiative heat transfer which is beyond the scope of this thesis.

- At the time the experiments were done, it was not possible to measure experimentally the temperature of the hot particle-gas mixture in the cavity under solar irradiation. The measured wall temperatures did not provide a useful means to estimate the particle temperature. Because

8.7. SUMMARY AND CONCLUSIONS

of this, the latter was extracted from A. Z'Graggen's model for the heat transfer inside the SynPet reactor featuring a 3D path length Monte-Carlo ray tracer.

- Key input parameters required to calculate the temperature profile in the cavity and the reactor performance are itself subjected to considerable errors: The solar power input Q_{solar} has an accuracy of $^{+9}_{-13}\%$ and the initial coke feed $\dot{n}_{C,0}$ ±5% [102].

- The model assumes a constant input of solar power. However, solar experiments are performed under transient conditions of the solar power input. A dynamic simulation of a solar experiment (outside the scope of this thesis) would require knowledge of the variation of power, temperature, and mass flow rates with time.

Chapter 9
Summary and outlook

This thesis was performed in the framework of a joint project of PDVSA, ETHZ and CIEMAT, aiming at the development of the required technology for the production of high quality syngas and hydrogen using extra-heavy Orinoco type of crude oil or derived residues such as petroleum coke using solar thermochemical processing.

The main objective of this research work is the experimental and theoretical analysis of the reaction kinetics and thermodynamics of the solar thermal gasification of petroleum coke. This process is a hybrid solar/fossil endothermic process in which the solid is used exclusively as the source for hydrogen production and solar power is used exclusively as the source for high temperature process heat.

Gasification of carbonaceous solids is a clean technology suitable to convert low or negative-value feed stocks such as residues from the processing of heavy and extra heavy crude oils into gaseous and liquid fuels. This process has recently gained importance for several reasons:

- In parallel to the depletion of fossil resources, the API gravity and the metal and sulfur content of the available feedstocks is increasing, making the use of nonconventional feedstocks such as high sulfur heavy oils, oil shales, tar, coke, and waste more important.

- The worldwide demand for refinery products and energy is increasing.

- Environmental restrictions for the combustion of carbonaceous fuels increase the importance of production processes for clean fuels. Gasification products are easily cleaned from contaminants and can serve as the basis material for high quality fuels.

The use of solar energy as a substitution of conventional internal combustion helps to preserve fossil resources. At the same time, the calorific value of the feedstock is upgraded and the release of pollutants to the environment is avoided. Solar gasification provides a viable transition path from todays fossil industry towards the long term goal of CO_2-free solar hydrogen production.

This thesis is part of the development of a new reactor technology for the solar thermal gasification of carbonaceous solids. The use of a state of the art high temperature entrained flow gasification reactor, as it is used in numerous conventional gasification facilities, extends traditional solar gasification concepts such as fluidized and packed bed reactors. High temperature entrained flow gasification is proven to be an efficient and reliable technology for the gasification of a large amount of different feedstocks and minimizes on the same time the discharge of pollutants to the environment.

9.1 Summary

A thermodynamic analysis of the steam gasification of petcoke is presented considering two types of petroleum coke, a Flexicoke and a delayed coke. Equilibrium computation of the stoichiometric system of petcoke and steam at 1 bar and 1300 K result in an equimolar mixture of H_2 and CO. A 2nd-law analysis for generating electricity using the gasification products indicates the potential of doubling the specific electrical output and halving the specific CO_2 emissions vis-à-vis conventional petcoke-fired power plants.

Rate laws for the overall gasification process including pyrolysis and reactive gasification are derived: Pyrolysis is modeled as a linear combination of first order decomposition reactions. The model for the reactive gasification is based on a Langmuir-Hinshelwood type reaction mechanism considering reversible sorption of gaseous species onto the carbon surface and irreversible reactions among adsorbed species and with molecules from the gas phase. A structural model for the porous solid (coke) and the surrounding gas film is used to account for mass transfer limitations.

The rate constants of the kinetic expressions were determined experimentally by dynamic thermogravimetry in the 900-1300 K temperature interval. Gasification experiments were performed in inert gas as well as Ar-H_2O-CO_2 mixtures with varying concentrations of the reactive gases. The gaseous products being H_2, CO, CO_2, CH_4, C_2H_2, C_2H_4, C_2H_6, H_2S, and COS were analyzed with a Varian Micro GC.

9.1. SUMMARY

The TG experiments further showed that the rate of the gasification reaction is significantly reduced once the samples are heated to temperatures higher than 1310 K, which is explained by thermal annealing and an experimentally validated loss of active surface and pore volume.

The effect of the particle size on the reaction rate is modeled with an empirical particle effectiveness expression according to $r_{obs} = \eta_P \, r_{intr}$, expressing the overall rate of the gasification reaction, r_{obs}, in terms of an effectivity constant and the intrinsic rate, r_{intr}.

Experiments with laboratory scale fluidized bed reactors are performed using two different modes of heat transfer to analyze possible effects on the reaction kinetics. A fluidized bed of coke particles is gasified using a quartz reactor and an opaque aluminum oxide reactor heated by concentrated thermal radiation from ETH's high-flux solar simulator. The observed gasification rates and the product composition are essentially the same for the two heat transfer modes, and consistent with the results from the TG experiments.

The thermodynamic and kinetic informations were used for the design of a 5 kW entrained flow prototype reactor. The so-called SynPet reactor consists of a 20 × 10 cm cylindrical cavity receiver that contains a 5 cm diameter opening closed with a quartz window to let in concentrated solar power. The reactor is operated with a fine ground PD coke slurry and has been successfully tested in the PSI solar furnace. Chemical conversion of carbon and steam reached up to 87% and 69%, respectively, for a solar power input in the range 3.3-6.6 kW and coke mass flow rate in the range 1.85-4.45 g/min. During the operation with a stoichiometric feed ($H_2O/C = 1$), the reactor produced a gas with $H_2/CO \approx 2$ and $CO_2/CO \approx 0.3$.

A mathematical model of the SynPet reactor is developed based on the axial dispersion model, using axial back mixing to match the residence time distribution of a non-ideal flow pattern. The axial dispersion number is obtained from residence time distribution measurements. The dispersion model is further extended by means of radial temperature profiles calculated from the experimental parameters of the SynPet campaign in the solar furnace. Kinetic expressions and rate data elaborated in Chapters 4 and 5 are used to calculate the rate of reaction. Mass transfer in the gas phase was analyzed and found to have no effect for the used particle size and the calculated temperatures.

Rate data obtained from TG experiments are corrected for the use with the process reactor in order to account for a large reactivity difference observed between coke samples gasified in the TG and those from the SynPet experiments. The higher reactivity of samples used in the process reactor is explained with the strong differences between the temperature history of the samples gasified in the SynPet reactor ($\tau \approx 1$ s) and those used in the TG ($\tau \approx 1$ h) and the DFB reactor ($\tau \approx 15$ min). Modeled data is compared with the solar experiments and showed reasonable accuracy with respect to gas composition and chemical conversion, keeping in mind the basic character of the model and the limitations associated with the temperature measurement.

In a subsequent experimental campaign [52, 101] it was demonstrated that the results obtained with dry petcoke feeding are reproducible using petcoke slurries as feedstock.

9.2 Outlook

Within the framework of this project, a second thesis at the Professorship in Renewable Energy Carriers at ETH Zurich is dedicated to the simulation of radiative heat transfer in chemically reacting systems in order to get a better understanding of the heat transfer mechanisms occurring in the SynPet reactor.

Ongoing experimental work with the 5 kW prototype reactor is dedicated to the implementation and testing of new feeder systems that are capable of handling petcoke slurries and liquid feedstocks including the direct use of vacuum residues mixed with water. Experiments with coke slurries have been performed successfully providing results that are rather similar to those presented in this thesis for steam/coke feeding. An additional experimental campaign with vacuum residues is currently running at PSI's solar furnace.

In parallel to the experiments with new feedstocks, a scale-up to a nominal thermal power of 500 kW is being performed. The new reactor has been designed by the ETH team and fabricated by CIEMAT. Figure 9.1 shows a photograph of the new reactor shell before assembling. The complete pilot chemical plant is being assembled at the Plataforma Solar de Almería, where the 500 kW reactor will be installed on top of a central receiver system, showed in Figure 9.2. The new reactor is scheduled for operation in 2007 and marks a further milestone in the development of a new gasification technology for the upgrading of refinery residues with concentrated solar energy.

As the reactor technology is demonstrated on the 500 kW scale, the ultimate objective is the construction of a solar gasification plant with multiple receiver/mirror fields for the gasification of residues on an industrial MW scale.

9.2. OUTLOOK

Figure 9.1: Reactor shell of the 500 kW scale-up before assembling.

Figure 9.2: Solar concentrating facility at the Plataforma Solar de Almería, Spain, where the testing of the 500 kW prototype chemical plant will take place.

List of Figures

2.1 Gasifier types for autothermic gasification: (a) fixed bed, (b) fluidized bed, and (c) entrained flow gasifier. 10

2.2 Schematic of the solar thermal gasification of petcoke. 13

2.3 SEM micrographs of PD coke particles at a magnification of 200× and 1000×, respectively. 22

2.4 SEM micrographs of Flexicoke particles. (a) shows the raw material, (b) are particles gasified in the DFB reactor with $X_C = 0.11$. Magnification: (a) 200×, (b) 500 ×. 22

3.1 Variation of the thermodynamic equilibrium composition with temperature of the system $CH_xO_y + (1-y)H_2O$ at 1 bar for Flexicoke. 25

3.2 Variation of the thermodynamic equilibrium composition with temperature of the system $CH_xO_y + (1-y)H_2O$ at 1 bar for PD coke. 25

3.3 Variation of the percent yield of H_2 and CO with temperature for the gasification of Flexicoke and PD coke, assuming the equilibrium composition in Figures 3.1 and 3.2, respectively. . 26

3.4 Enthalpy change of reaction (3.2) as a function of temperature T for the two types of petcoke, when the reactants are fed at 300 K and the products are obtained at T having the equilibrium composition given in Figures 3.1 and 3.2. 26

3.5 Flow sheet diagram used for the 2nd-law analysis. 29

4.1 Progressive-conversion and shrinking unreacted core model. . . 35

4.2 Schematic of a spherical particle with concentration gradient and spherical shell of thickness ΔR. 37

4.3 Schematic of the grain model. The coke particle is assumed to be an aggregation of smaller grains. The grains are modeled as spheres of uniform size, each of which reacts with steam according to the SCM model. 38

4.4 Schematic of the mass transfer model for the gas phase: Steam is transfered from the bulk gas to the surface of a spherical petcoke particle, where it reacts with carbon to form H_2, CO, and CO_2. 42

5.1 Cross-sectional schematics of the Netzsch steam furnace and high temperature furnace. 56

5.2 Empty crucible and crucible with mounted coke sample. 56

5.3 TG and DTG data for the pyrolysis of Flexicoke and PD coke. 62

5.4 Weight loss and product gas composition as a function of temperature for the pyrolysis of PD coke. 62

5.5 Weight loss and product gas composition as a function of temperature for the steam gasification of Flexicoke. 64

5.6 Weight loss and product gas composition as a function of temperature for the steam gasification of PD coke. 64

5.7 Weight loss as a function of temperature for the gasification of Flexicoke in binary H_2O-Ar and CO_2-Ar mixtures. 66

5.8 Weight loss as a function of temperature for the gasification of PD coke in binary H_2O-Ar and CO_2-Ar mixtures. 66

5.9 Weight loss as a function of temperature for the steam gasification of Flexicoke. 67

5.10 Weight loss as a function of temperature for the steam gasification of PD coke. 67

5.11 Calculation of r_{intr} for Flexicoke via extrapolation of isothermal rate data to $R_P = 0$. 69

5.12 Calculation of r_{intr} for PD coke via extrapolation of isothermal rate data to $R_P = 0$. 69

5.13 Experimental particle effectiveness factor η_P for the steam gasification of Flexicoke and PD coke. 70

5.14 Thermal deactivation of PD coke above 1300 K during gasification experiments with different reactive gas compositions. . 72

5.15 Weight loss as a function of temperature for the gasification of thermally pretreated PD coke. 73

LIST OF FIGURES

5.16 Rate of reaction as a function of temperature for the gasification of thermally pretreated PD coke. 73

5.17 Experimental error (noise) of the TG during the pyrolysis and gasification runs 75

5.18 Measured and modeled conversion rate dX_P/dt obtained for the pyrolysis of Flexicoke 76

5.19 Measured and modeled conversion rate dX_P/dt obtained for the pyrolysis of PD coke. 76

5.20 Experimentally measured and modeled reaction rate as a function of temperature for the gasification of Flexicoke and PD coke. 80

5.21 Reaction rate, $-r_C$ (s^{-1}), as a function of the H_2O and CO_2 partial pressures at 1273 K for the gasification of Flexicoke and PD coke. 81

6.1 Flowchart of the fluidized bed reactor setup in the ETH's high-flux solar simulator. 86

6.2 Quartz fluidized bed reactor before and after experiment. ... 86

6.3 The ETH's solar simulator with the quartz reactor during an experiment. 86

6.4 Fluidized bed temperature and product gas composition during a representative experimental run with Flexicoke and 10% H_2O-Ar. 88

6.5 Steam and carbon conversion for the experimental run of Figure 6.4 with Flexicoke and 10% H_2O-Ar. 88

6.6 Product gas composition and steam conversion as a function of the fluidized bed temperature for Flexicoke gasification using a feed gas of 10% H_2O-Ar. 90

6.7 Product gas composition and steam conversion as a function of the fluidized bed temperature for PD coke gasification using a feed gas of 10% H_2O-Ar. 90

6.8 Product gas relative composition and steam conversion as a function of temperature for the steam gasification of Flexicoke and PD coke with direct irradiation of the fluidized bed. Argon and excess steam are omitted. 91

6.9 Variation of H_2S concentration in the product gas with temperature for the steam gasification of both petcokes. 91

6.10 Reactor temperature and product gas concentration during a *dry* experimental run with Flexicoke. 93

6.11 Variation of the elementary composition of PD coke with temperature. 93

6.12 Variation of the specific surface and micropore area of PD coke with temperature. 94

6.13 Variation of the micropore volume and the average pore diameter of PD coke with temperature. 94

6.14 Fluidized bed reactor setup for the use of opaque ceramics fluidization tubes with radiation shield and quartz dome. . . . 96

6.15 Bed temperature as a function of the arc current for different fluidization tubes. 97

6.16 Fluidized bed temperature and product gas composition during a representative experimental run with PD coke in 10% H_2O-Ar. 99

6.17 Steam and carbon conversion for the experimental run of Figure 6.16 with PD coke and 10% H_2O-Ar. 99

6.18 Product gas composition and steam conversion as a function of the fluidized bed temperature for the gasification of Flexicoke using 10% H_2O-Ar. 100

6.19 Product gas composition and steam conversion as a function of the fluidized bed temperature for the gasification of PD coke using 10% H_2O-Ar. 100

6.20 Product gas relative composition and steam conversion as a function of temperature for the steam gasification of Flexicoke, and PD coke with indirect irradiation of the fluidized bed. . . . 102

6.21 Experimentally measured vs. modeled product flow rates. . . . 104

6.22 Concentration profile along the dimensionless reactor axis and boundary conditions for the steam gasification of PD coke in 10% H_2O-Ar at 1572 K. 105

6.23 Arrhenius plots of K_1a. 106

6.24 Arrhenius plots of K_2a. 107

6.25 Arrhenius plots of K_3. 107

6.26 Temperature dependence of the H_2/CO and CO_2/CO molar ratio in the product gas for the experiments with the DFB and IFB reactor. 109

6.27 Comparison of the carbon consumption rate, r_C for the steam gasification of Flexicoke and PD coke using the TG as well as the DFB and IFB setups. 111

7.1 Scheme of solar chemical reactor configuration for the steam gasification of petcoke. 116
7.2 Schematic of the experimental setup at PSI's solar furnace. . . 118
7.3 Off-gas composition of the reference experiment measured by GC and IR spectroscopy for the steam gasification of PD coke in the SynPet reactor. 120
7.4 Conversion of carbon and steam calculated with Equations (7.1) and (7.2), respectively, for the experiment in Figure 7.3. 121
7.5 Pyrometer temperature, $T_{reactor}$, wall temperature, $T_{inconell}$, off- gas temperature, $T_{gas,out}$, and thermal input, Q_{solar}, of the experiment in Figure 7.3. 121
7.6 Steam and carbon conversion as function of the pyrometer temperature (cf. Figure 7.5) for the SynPet experiments with PD coke. 123
7.7 Steam and carbon conversion as function of the coke and steam feed rates for the SynPet experiments with PD coke. 123
7.8 Steam and carbon conversion as function of the H_2O/C molar ratio in the feed for the SynPet experiments with PD coke. . . 124
7.9 Product composition as function of the pyrometer temperature for the SynPet experiments with PD coke. 126
7.10 Product composition as function of the coke and steam feed rates for the SynPet experiments with PD coke. 126
7.11 Product composition as function of the H_2O/C molar ratio in the feed for the SynPet experiments with PD coke. 127

8.1 Map showing which flow model should be used. 132
8.2 One-dimensional grid point cluster for the discretization. . . . 136
8.3 Discretization scheme of the reactor with the boundary elements 1 and N. 138
8.4 Flowchart of the experimental setup for the measurement of the RTD. 141
8.5 Experimentally determined residence time distribution of the SynPet reactor (E and F curves). 141

8.6 TG curves of PD coke residues collected after three runs in the solar furnace compared to the standard sample. 144

8.7 Rate of steam gasification with $p_{H_2O} = 0.6$ bar for three PD coke samples collected in the cyclone after the SynPet reactor. 145

8.8 Gas composition and carbon conversion axial profiles for the SynPet reactor operated at 1800 K with feed flow rates of the experiment presented in Figure 7.3 for four different particle sizes. 147

8.9 Radial distribution of the particle temperature for the conditions of the SynPet experiments in the solar furnace (cf. Chapter 7). The data are calculated with the Monte-Carlo radiation model [100]. 152

8.10 Peak temperature, integral mean temperature, and wall temperature of the coke particles extracted from Figure 8.9. The experimental runs are ordered by increasing pyrometer temperature. 152

8.11 Schematic showing the radial discretization of the SynPet reactor with respect to temperature. 153

8.12 Radial variation of carbon conversion, residence time and temperature at the reactor outlet, calculated for the reference experiment in Chapter 7. 154

8.13 Radial variation of the carbon conversion at different positions along the reactor axis for the reference experiment. 154

8.14 Radial profiles of the gas composition, y_i, and the carbon conversion, X_C, of the reference experiment at the reactor outlet ($L = L_0$). 156

8.15 Axial profiles of the gas composition, y_i, and the carbon conversion, X_C, of the reference experiment at $R/R_0 \approx 0.6$ having a calculated temperature of 1800 K. 156

8.16 Experimental and simulated off-gas composition, y_i, for the SynPet experiments in the solar furnace (cf. Chapter 7). . . . 158

8.17 Experimental and simulated carbon conversion, X_C, for the SynPet experiments in the solar furnace (cf. Chapter 7). . . . 160

8.18 Experimental and simulated steam conversion, X_{H_2O}, for the SynPet experiments in the solar furnace (cf. Chapter 7). . . . 160

9.1 Reactor shell of the 500 kW scale-up before assembling. 169

9.2 Solar concentrating facility at the Plataforma Solar de Almería, Spain, where the testing of the 500 kW prototype chemical plant will take place. 169

List of Tables

2.1	Proximate analysis for Flexicoke and PD coke.	19
2.2	Approximate main elemental chemical composition (ultimate analysis), low heating value, and elemental molar ratios of H/C and O/C, for PD coke and Flexicoke.	19
2.3	Coke samples used for the experiments.	21
3.1	Exergy analysis of the solar steam gasification of petcoke using the process modeling shown in Figure 3.5.	30
5.1	Analyzed gas mixtures used for the GC calibration	57
5.2	Pyrolysis runs in the high temperature furnace with the corresponding experimental parameters.	60
5.3	Reactive gasification runs in the steam furnace with the corresponding experimental parameters.	60
5.4	Numerical values for the parameters a, b, c and d in Equation (5.4)	71
5.5	Kinetic parameters for the pyrolysis of petroleum coke.	77
5.6	Arrhenius parameters of the rate constants derived by applying the oxygen-exchange mechanism for the H_2O-CO_2 gasification of PD coke.	78
5.7	Arrhenius parameters of the rate constants derived by applying the extended mechanism for the H_2O-CO_2 gasification of Flexicoke and PD coke.	79
6.1	Mean properties of petcoke particles used as reactants in this study.	85
6.2	Experimental parameters of hydrogen and oxygen released by pyrolysis during *dry* experimental runs.	92

6.3 Characteristics of the fluidization tubes tested with the setup in Figure 6.14. 96

6.4 Arrhenius parameters for the rate law of steam gasification of Flexicoke and PD coke respectively, measured with the DFB reactor (Chapter 6.1). 108

6.5 Arrhenius parameters for the rate law of steam gasification of Flexicoke and PD coke respectively, measured with the IFB reactor (Chapter 6.2). 108

8.1 Distribution parameters and vessel dispersion number for the SynPet reactor. 141

8.2 Numerical values for the parameters in Equation (8.56) obtained by minimization of the error between the left hand side and the right hand side of Equation (8.55). 145

Bibliography

[1] Y. K. Ahn and W. H. Fischer. Production of hydrogen from coal and petroleum coke: thechnical and economic perspectives. *International Journal of Hydrogen Energy*, 11(12):783–788, 1986.

[2] R. Alvarez. *Solar thermal steam gasification of petcoke in an opaque fluidized bed reactor*. Master thesis, Swiss Federal Institute of Technology (ETH), 2004.

[3] V. Anathakrishnan, W. N. Gill, and A. J. Barduhn. Laminar dispersion in capillaries. I. mathematical analysis. *AIChE Journal*, 11(6):1063–1072, 1965.

[4] M. J. Antal, L. Hofmann, J. R. Moreira, C. T. Brown, and R. Steenblik. Design and operation of a solar fired biomass flash pyrolysis reactor. *Solar Energy*, 30(4):299–312, 1983.

[5] D. B. Anthony and J. B. Howard. Coal devolatilization and hydrogasification. *AIChE Journal*, 22(4):625–656, 1967.

[6] W. H. Beattie, R. Berjoan, and J. P. Coutures. High-temperature solar pyrolys of coal. *Solar Energy*, 31(2):137–143, 1983.

[7] A. Belghit and M. Daguenet. Study of heat and mass transfer in a chemical moving bed reactor for gasification of carbon using an external radiative source. *Int. J. Heat Mass Transfer*, 32(11):2015–2025, 1989.

[8] A. Belghit, M. Daguenet, and A. Reddy. Heat and mass transfer in a high temperature packed moving bed subject to an external radiative source. *Chemical Engineering Science*, 55:3967–3978, 2000.

[9] R. Berber and E. A. Fletcher. Extracting oil from shale using solar energy. *Energy*, 13(1):13–23, 1988.

[10] R. Bertocchi, J. Karni, and A. Kribus. Experimental evaluation of a non-isothermal high temperature solar particle receiver. *Energy*, 29(5-6):687–700, 2004.

[11] R. B. Bird, W. E. Stewart, and E. N. Lightfoot. *Transport phenomena*. John Wiley & Sons, New York, 2nd edition, 2002.

[12] J. D. Blackwood and A. J. Ingeme. The reaction of carbon with carbon dioxide at high pressure. *Australian Journal of Chemistry*, 13:194–209, 1960.

[13] J. D. Blackwood and F. McGrory. The carbon-steam reaction at high pressure. *Australian Journal of Chemistry*, 11:16–33, 1957.

[14] M. Bohnet, editor. *Ullmann's encyclopedia of industrial chemistry*. Wiley-VCH, Weinheim, 2003.

[15] R. W. Bryers. Utilization of petroleum coke and petroleum coke/coal blends as a means of steam rizing. *Fuel Processing Technology*, 44(1-3):121–141, 1995.

[16] E. L. Cussler. *Diffusion: mass transfer in fluid systems*. Cambridge University Press, Cambridge etc., 2nd edition, 1997.

[17] J. K. Dahl, B. Karen, F. Ryan, T. Stanislaus, A. W. Weimer, A. Lewandowski, C. Binham, A. Smeets, and A. Schneider. Rapid sola-thermal dissociation of natural gas in an aerosol flow reactor. *Energy*, 29(5-6):715–725, 2004.

[18] R. DiPanfilo and N. O. Egiebor. Activated carbon production from synthetic crude coke. *Fuel Processing Technology*, 46:157–169, 1996.

[19] E. L. Doering and G. A. Cremer. The SHELL coal gasification process: the demkolec project and beyond. In *Power-Gen Americas '93*, Dallas TX, 1993.

[20] E. L. Doering and G. A. Cremer. Advances in the shell coal gasification process. *Preprints of Papers - American Chemical Society, Division of Fuel Chemistry*, 40(2):312–317, 1995.

[21] S. Ergun. Kinetics of the reactions of carbon dioxide and steam with coke. Technical report, U.S. Bureau of Mines, 1962.

[22] M. Fasciana and F. Noembrini. *Solar petcoke gasification - direct irradiation*. Master thesis, Swiss Federal Institute of Technology (ETH), 2003.

[23] J. L. Figueiredo, J. A. Moulijn, and (editors). Carbon and coal gasification: science and technology. In *Carbon and coal gasification: science and technology*, Alvor, 1986.

[24] M. Flechsenhar and C. Sasse. Solar gasification of biomass using oil shale and coal as candidate materials. *Energy*, 20(8):803–810, 1995.

[25] E. Furimsky. Gasification in petroleum refinery of 21st century. *Oil & Gas Science and Technology*, 54(5):597–618, 1999.

[26] J. Gadsby, F. J. Long, P. Sleightholm, and K. W. Sykes. The mechanism of the carbon dioxide-carbon reaction. *Proc. Roy. Soc. (London)*, A193:357–376, 1948.

[27] J. H. Gary and G. E. Handwerk. *Petroleum refining*. Marcel Dekker, New York, 4th edition, 2001.

[28] A. Gómez-Barea, P. Ollero, and R. Arjona. Reaction-diffusion model of TGA gasification experiments for estimating diffusional effects. *Fuel*, 84(12-13):1695–1704, 2005.

[29] D. W. Gregg. Apparatus and method for solar coal gasification. Patent US 4'229'184, 1980.

[30] D. W. Gregg, W. R. Aiman, H. H. Otsuki, and C. B. Thorsness. Solar coal gasification. *Solar Energy*, 24(3):313–321, 1980.

[31] D. W. Gregg, R. W. Taylor, J. H. Campbell, J. R. Taylor, and A. Cotton. Solar gasification of coal, activated carbon, coke and coal and biomass mixtures. *Solar Energy*, 25(4):353–364, 1980.

[32] P. Haueter, T. Seitz, and A. Steinfeld. A new high-flux solar furnace for high-temperature thermochemical research. *Jounal for Solar Energy Engineering*, 121(1):77–80, 1999.

[33] W. F. Hemminger and H. K. Cammenga. *Methoden der thermischen Analyse*, volume XXIV of *Anleitungen für die chemische Laboratoriumspraxis*. Springer-Verlag, Berlin, 1989.

[34] D. Hirsch and A. Steinfeld. Radiative transfer in a solar chemical reactor for the co-production of hydrogen and carbon by thermal decomposition of methane. *Chemical Engineering Science*, 59(24):5771–5778, 2004.

[35] D. Hirsch and A. Steinfeld. Solar hydrogen production by thermal decomposition of natural gas using a vortex-flow reactor. *International Journal of Hydrogen Energy*, 29(1):47–55, 2004.

[36] D. Hirsch, P. v. Zedtwitz, T. Osinga, J. Kinamore, and A. Steinfeld. A new 75 kW high-flux solar simulator for high-temperature thermal and thermochemical research. *ASME-Journal of Solar Engineering*, 125(1):117–120, 2003.

[37] The MathWorks Inc. Matlab R14, 2005.

[38] G. Ingel, M. Levy, and J. M. Gordon. Oil-shale gasification by concentrated sunlight: An open-loop solar chemical heat pipe. *Energy*, 17(12):1189–1197, 1992.

[39] H. Kabs. Operational experience with Siemens-Westinghouse SOFC cogeneration systems. In *Fuel Cell Home Conference*, pages 197–209, Switzerland: Luzern, 2001.

[40] S. Kajitani, S. Hara, and H. Matsuda. Gasification rate analysis of coal char with a pressurized drop tube furnace. *Fuel*, 81(5):539–546, 2002.

[41] A. Karcz and S. Porada. Formation of C_1-C_3 hydrocarbons during pressure pyrolysis and hydrogasification in relation to structural changes in coal. *Fuel*, 76(6):806–809, 1995.

[42] R. E. Kirk and D. F. Othmer. *Concise encyclopedia of chemical technology*. Wiley, New York, 1999.

[43] D. Kocaefe, A. Charette, and L. Castonguay. Green coke pyrolysis: investigation of simultaneous changes in gas and solid phases. *Fuel*, 74(6):791–799, 1995.

[44] T. Kodoma, Y. Kondoh, T. Tamagawa, A. Funatoh, K. I. Shimizu, and Y. Kitayama. Fluidized bed coal gasification with CO_2 under direct irradiation with concentrated visible light. *Energy & Fuels*, 16(5):1264–1270, 2002.

[45] S. Kraeupl and A. Steinfeld. Operational performance of a 5 kW solar chemical reactor for the co-production of zinc and syngas. *Journal of Solar Energy Engineering*, 125(1):124–126, 2003.

[46] H. Kubiak and H. Lohner. Study relating to the use of solar energy for the allothermal gasification of coal. *Sol. Therm. Energy Util.*, 6:327–340, 1992.

[47] N. M. Laurendeau. Heterogenous kinetics of coal char gasification and combustion. *Prog. Energy Combust. Sci*, 4(4):221–270, 1978.

[48] S. Lee, J. C. Angus, R. V. Edwards, and N. C. Gardner. Noncatalytic coal char gasification. *AIChE Journal*, 30(4):583–593, 1984.

[49] O. Levenspiel. *Chemical reaction engineering*. Wiley, New York, 3rd edition, 1999.

[50] F. J. Long and K. W. Sykes. The mechanism of the steam-carbon reaction. *Proc. Roy. Soc. (London)*, A193:377–399, 1948.

[51] H. H. Lowry and M. A. Elliott, editors. *Chemistry of coal utilization*. Wiley, New York, 1981.

[52] G. Maag. *Solar thermal gasification of petcoke*. Semester thesis, Swiss Federal Institute of Technology (ETHZ), 2006.

[53] W. Malburg and H. Treiber. Solar reactor. Patent DE 2836179, 1980.

[54] V. K. Mathur, R. W. Breault, and S. M. Lakshmanan. Synthesis gas production by the solar gas process. *Energy Progress*, 2(2):91–95, 1982.

[55] V. K. Mathur, S. Lakshmanan, F. K. Manasse, V. Venkataramanan, and R. W. Breault. A continuous two stage solar coal gasification system. *AIChE Symposium Series*, 77(210):47–54, 1981.

[56] J. Matsunami, S. Yoshida, Y. Oku, O. Yokota, O. Yokota, Y. Tamaura, and M. Kitamura. Coal gasification with CO_2 in molten salt for solar thermal/chemical energy conversion. *Energy*, 25:71–79, 2000.

[57] M. Mentser and S. Ergun. A study of the carbon dioxide-carbon reaction by oxygen exchange. Technical report, U.S. Bureau of Mines, 1973.

[58] M. Meyer and R. Rusch. *Optimierung von Reaktor-Strömung und Verweilzeit*. Semester thesis, ETH, 2004.

[59] A. Morales, Y. Araujo, and E. Lima. In *International Forum on Technology for Heavy and Extra-Heavy Oil*, Puerto La Cruz, Venezuela, 2003.

[60] H. J. Mühlen. *Zum Einfluss der Produktgase auf die Kinetik der Wasserdampfvergasung in Abhängigkeit von Druck und Temperatur*. PhD thesis, Universität Essen (D), 1983.

[61] H. J. Mühlen. High temperature, high pressure thermogravimetry of coal gasification - apparatus, data acquisition and numerical evaluation. *Thermochimica Acta*, 103(1):163–168, 1986.

[62] H. J. Mühlen, K. H. v. Heek, and H. Jüntgen. Kinetic studies of the steam gasification of char in the presence of H_2, CO_2 and CO. *Fuel*, 64(7):944–949, 1985.

[63] R. Müller, P. v. Zedtwitz, A. Wokaun, and A. Steinfeld. Kinetic investigation on steam gasification of charcoal under direct high flux irradiation. *Chemical Engineering Science*, 58(22):5111–5119, 2003.

[64] J. P. Murray and E. A. Fletcher. Reaction of steam with cellulose in a fluidized bed using concentrated sunlight. *Energy*, 19(10):1083–1098, 1993.

[65] P. Ollero, A. Serrera, R. Arjona, and S. Alcantarilla. Diffusional effects in TGA gasification experiments for kinetic determination. *Fuel*, 81(15):1899–2017, 2002.

[66] S. V. Patankar. *Numerical heat transfer and fluid flow*. Series in Comptutational and physical processes in mechanics and thermal sciences. Taylor & Francis, Oxford, 1980.

[67] G. J. Pitt. The kinetics of the evolution of volatile products from coal. *Fuel*, 41:267–274, 1962.

[68] B. E. Poling, J. M. Prausnitz, and J. P. O'Connell. *The properties of gases and liquids*. McGraw-Hill, New York, 5 edition, 2001.

[69] A. S. Qader. Solar heated fluidized bed gasification system. Patent US 150115, 1980.

[70] A. E. Reif. The mechanism of the carbon dioxide-carbon reaction. *J Phys Chem*, 56:785–788, 1952.

[71] J. Rezaiyan and N. P. Cheremisinoff. *Gasification technologies.* Taylor & Francis, Boca Raton FL, 2005.

[72] A. Roine. Outokumpu HSC chemistry for Windows, 1997.

[73] D. Ruthven, S. Farooq, and K. S. Knaebel. *Pressure swing adsorption.* Wiley, New York, 1993.

[74] D. H. Sauter. Coal. In B. Elvers and S. Hawkins, editors, *Ullmann's encyclopedia of industrial chemistry*, volume A7, pages 153–197. VCH Verlagsgesellschaft, Weinheim, Germany, 1986.

[75] R. H. Schlosberg, editor. *Chemistry of coal conversion.* Plenum Press, New York, London, 1985.

[76] V. Seebauer, J. Petek, and G. Staudinger. Effects of particle size, heating rate and pressure on measurement of pyrolysis kinetics by thermogravimetric analysis. *Fuel*, 76(13):1277–1282, 1997.

[77] R. Siegel and J. Howell. *Thermal radiation heat transfer.* Taylor & Francis, New York, 4th edition, 2002.

[78] J. M. Smith. *Chemical engineering kinetics.* McGraw-Hill, New York, 3rd edition, 1981.

[79] J. G. Speight and B. Özüm. *Petroleum refining processes.* Marcel Dekker, Inc, New York, 2002.

[80] W. A. Stahel. *Statistische Datenanalyse.* Vieweg, Braunschweig (D), 2nd edition, 1999.

[81] A. Steinfeld. Solar hydrogen production via a 2-step water splitting thermochemical cycle based on Zn/ZnO redox reactions. *International Journal of Hydrogen Energy*, 27(6):611–619, 2002.

[82] A. Steinfeld. Solar thermochemical production of hydrogen - a review. *Solar Energy*, 78(5):603–615, 2005.

[83] A. Steinfeld, M. Brack, A. Meier, A. Weidenkaff, and D. Wuillemin. A solar chemical reactor for co-production of zinc and synthesis gas. *Energy*, 23(10):803–814, 1998.

[84] A. Steinfeld and A. Meier. Solar fuels and materials. In C. J. Cleveland, editor, *Encyclopedial of Energy*, volume 5, pages 623–637. Elsevier Academic Press, Amsterdam, 2004.

[85] A. Steinfeld and R. Palumbo. Solar thermochemical process technology. In R. A. Meyers, editor, *Encyclopedia of Physical Science and Technology*, volume 15, pages 237–256. Academic Press, San Diego CA, 3rd edition, 2001.

[86] G. J. Stiegel and R. C. Maxwell. Gasification technologies: the path to clean, affordable energy in the 21st century. *Fuel Processing Technology*, 71:79–97, 2001.

[87] K. M. Sundaram. Catalyst effectiveness factor for Langmuir-Hinshelwood-Hougen-Watson kinetic expressions. *Chemical Engineering Communications*, 15(5-6):305–311, 1982.

[88] J. Szekely and J. W. Evans. A structural model for gas-solid reacttions with a moving boundary. *Chemical Engineering Science*, 25(6):1091–1107, 1970.

[89] J. Szekely and J. W. Evans. A structural model for gas-solid reactions with a moving boundary - II. *Chemical Engineering Science*, 26(4):1901–1913, 1971.

[90] R. W. Taylor, R. Berjoan, and J. P. Coutures. Solar gasification of carbonaceous materials. *Solar Energy*, 30(6):513–525, 1983.

[91] H. R. Tschudi and G. Morian. Pyrometric temperature measurements in solar furnaces. *Jornal of Solar Energy Engineering*, 123(2):164–170, 2001.

[92] C. Ulloa, A. L. Gordon, and X. García. Distribution of activation energy model applied to the rapid pyrolysis of coal blends. *Journal of Analytical and Applied Pyrolysis*, 71(2):465–483, 2004.

[93] E. T. v. d. Laan. Notes on the diffusion-type model for the longitudinal mixing in flow. *Chemical Engineering Science*, 7(3):187–191, 1958.

[94] P. v. Zedtwitz, J. Petrasch, D. Trommer, and A. Steinfeld. Hydrogen production via the solar thermal decarbonization of fossil fuels. *Solar Energy*, 2005.

[95] P. v. Zedtwitz and A. Steinfeld. The solar thermal gasification of coal - energy conversion efficiency and CO_2 mitigation potential. *Energy*, 28(5):441–456, 2003.

[96] D. Vamvuka. Gasification of coal. *Energy Exploration and Exploitation*, 17(6):515–581, 1999.

[97] S. M. Walas. *Reaction kinetics for chemical engineers*. Butterworths, Boston, 1989.

[98] J. R. Welty. *Fundamentals of momentum, heat and mass transfer*. Wiley, New York, 4 edition, 2001.

[99] C. Y. Wen. Noncatalytic heterogeneous solid fluid reaction models. *Industrial & Engineering Chemistry*, 60(9):34–54, 1968.

[100] A. Z'Graggen. Static Monte-Carlo simulation of the SynPet solar reactor. personal communication, 2006.

[101] A. Z'Graggen, P. Haueter, G. Maag, A. Vidal, and M. Romero. Hydrogen production by steam gasification of petroleum coke using concentrated solar power - reactor experimentation with slurry feeding. In R. Campbell-Howe, editor, *Solar 2006*, Denver, CO, 2006.

[102] A. Z'Graggen, P. Haueter, D. Trommer, M. Romero, J. C. d. Jesus, and A. Steinfeld. Hydrogen production by steam-gasification of petroleum coke using concentrated solar power - II. reactor design, testing and modeling. *International Journal of Hydrogen Energy*, 31(6):797–811, 2006.

[103] A. Z'Graggen and A. Steinfeld. A two-phase reactor model for the steam gasification of carbonaceous materials under concentrated thermal radiation. *Chemical Engineering & Processing*, submitted, 2007.